The Religion of Science

The Big Picture

by N. Lee Swanson

Religion, n.
3. (a) any specific system of belief, worship, conduct, etc., often involving a code of ethics and a philosophy; (b) loosely, any system of beliefs, practices, ethical values, etc. resembling, suggestive of, or likened to such a system. --Webster's New Universal Unabridged Dictionary, 2nd edition, Simon and Schuster, NY, 1972 (3rd defn.)

ISBN-13: 978-1475058598
ISBN-10: 1475058594

CreateSpace.com

Acknowledgments

This book was not written without a considerable amount of encouragement, feedback and inspiration. The following people gave me one or all of the above. I am exceedingly grateful to Alan, Francisco, Fred, Donna, Heather, Lindy, Jim, Gayle, Skip, Christian and Chiyomi. You know who you are. Thank you.

The Religion of Science

Author's Note

All scientific books and papers are written in third person passive voice. This is because scientists believe they are unbiased observers. They want you to believe they are objective. This illusion is easier to maintain if they never use the word "I". If absolutely necessary, they will use the word "we" (even if they were alone when the thing was observed, as in the royal "we").

To write scientific reports and conclusions in the third person passive voice is cumbersome and inelegant, and often results in sentences implying that some instrument, like a computer or an oscilloscope did it. Because of this, I wrote a report in first person at my first professional place of employment. As a result my boss recommended me for a scientific writing course. He thought I didn't know the rules. In this book, I am choosing to use whichever voice seems to make the most sense. I like first person active because it's clean, honest and it lets the reader know that I take responsibility for my work and conclusions.

I am now semi-retired, which gives me the freedom to write this book. I could not have written it while I was actively practicing traditional science because I would have suffered the consequences. I would have incurred the ire and wrath of my colleagues, possibly losing my employment, surely losing my research funding and probably not getting my papers published. This is the punishment for disagreeing with the scientific priesthood. I was not really able to think about the underpinnings of science until I was no longer a practicing scientist. I am now free to express my thoughts. Hopefully, I won't be burned at the stake.

I have earned B.S., M.S. and PhD. degrees in physics. I have over 25 years experience as a scientist working in a government laboratory, teaching at the university, and private consulting. I mention this only to inform the reader that, having been an insider, I have some authority to comment on this subject.

This book is intended for the general audience, though some background in science would be helpful. I have tried to define terms that might be unknown to the general reader. When I have thought it necessary to expound in more technical detail, I have placed that information in a box at the end of the section or chapter. Those who are interested in the detail can read the information in the box. I have tried to keep equations to a minimum. Those who are not interested can skip the boxes without losing the thread.

Throughout this book I endeavor to clearly differentiate between what is actually known (measured), what is inferred from what is known (conjecture), and what is purely opinion, mine or anyone else's.

I ask of you, the reader, to think. I also ask that you temporarily suspend your ideas about science and about what is and is not possible. After you finish this book (or throw it against the wall), you can always adopt them again. I am *not* asking you to suspend your reason. On the contrary, I ask that you seriously employ your reasoning capacity.

What I say in this book is what I believe to be true at this time. Tomorrow I may think differently because tomorrow I will be a different person. Also, what is true today may not be true tomorrow. There may not be such a thing as Truth, with a capital T. But that should not stop us from looking. And here's the important thing: once we are convinced we've found the Truth, we stop looking. If we convince ourselves that Truth can't be found, we don't look at all. It's vital that we keep looking. Because figuring out why we're here, who and what we are is the only game there is.

The Religion of Science

Introduction

I hope to show, through careful reason and logic, that there is no such thing as a fact. There are limits to what we can know and nothing that we can prove. Furthermore, because of blind adherence to facts, we have completely misunderstood how our world works. No doubt we still misunderstand.

Scientists and mathematicians have given us many great and wonderful ideas, not to mention very useful gadgets. But success is no measure of correctness, as the geocentric model of the solar system (to which we shall return) attests.

The central theme of this book is this: Don't believe anything that anyone else tells you. Including me. Find out for yourself. Only your direct experience, both internal and external, matters. Put your faith in yourself. Act on your own authority. If you unquestioningly believe what the priests of *any* religion tell you, then you are giving up control of your own life. It's time to think for yourself. You are the master of your destiny.

It seems that we have been plunked here on this planet that we have named Earth without any clues as to where we are, why we are here, or even who or what we are. We have busily set about to try to orient ourselves by searching for answers to these questions in various ways.

In our current day it is common to treat information that comes from within with suspicion and mistrust. The only information considered reliable is obtained from without, through direct interaction with the Earth and Its constituent parts. We set about to make up laws governing the motion of baseballs and such and then test our laws "out there" in the world. We have somehow evolved to the place where no other method will do. We have put our faith in the scientific priests who make up the laws.

Notice that none of the basic questions (where/why/who are we?) are answered by scientific inquiry. Science is only useful for possibly answering, "How does it work?" The other questions have been relegated to religion or philosophy although, historically, religion and philosophy were one and the same.

The adherents of science believe their method to be clear-cut and superior to any other. They claim the intellectual (even the moral) high ground, denigrating any who dare to disagree with either their methods or their results. Those who question are presumed lacking in sense.

In order to find out how our world works, scientists must begin with some basic assumptions about what is so (reality). Then they must decide, within that framework of assumptions, what is important or most fundamental to the workings of reality. They then work out their theories as if the assumptions and fundamentals are true. I have no quarrel thus far. One has to start somewhere after all.

The trouble happens because the scientific adherents forget, or were never told, that assumptions were made in the first place. If they think about it at all, they think the initial assumptions are obviously what is so.

There are signs that something is seriously wrong with the picture we currently hold of our world. In an effort to uncover the problems, I begin in Part I with our cultural and personal beliefs along with a brief history of how we got them. I then resort to philosophy to show that we can know nothing for sure. In Part II, I outline some of the problems with physics as it is currently understood and why we need a new paradigm. In Part III I offer some ideas about how we might begin with a new paradigm. I hope you enjoy this journey.

Part I

Core Beliefs & How We Got Them

1. Frameworks and Beliefs

"Lord girl, there's only two or three things I know for
sure. Only two or three things. That's right. Of course, it's
never the same things, and I'm never as sure as I'd like to be."
--Dorothy Allison, **Two or Three Things I Know for Sure**, Penguin,
1995.

Any person born into any given culture or society
is taught to accept the basic premises of that society.
This is necessary in order to communicate effectively
with our fellows. To question societal norms is
considered misguided at best, crazy at worst. We are
trained as very young children to distinguish between
what is and isn't real (*i.e.* what is or isn't perceived by
other human beings). We are told that, "There are no
monsters under the bed," and "Your playmate is
imaginary." As children, most of us were led to believe
there is a God, a Heaven and a Hell. Most adults would
agree that God, Heaven and Hell are beliefs requiring
faith, rather than facts since the existence of these
cannot be proved. And most people would say it is a
fact that there are no monsters under the bed. If we
refuse to conform to accepted facts then we are indeed
considered crazy.

We commonly associate facts with proof and
beliefs with faith. The question is: what is fact and what

is belief? At first glance this may sound like a straight-forward question, but this entire book is inspired by this question.

There is probably nothing that comes closer to hocus-pocus than the medicine of yesteryear. As a case in point, consider King Charles II of England. In 1865 he suffered a stroke. The royal physicians first drained twp cups of blood from Charles. Seeing no improvement, they gave him an enema. When he still showed no improvement, they gave him a dose of sneezing powder, rubbed pigeon dung on his feet, seared his shaved head and finally put a red hot iron on his feet. Mercifully for poor Charles, he died within four days of receiving the Royal Treatment.[1]

As ridiculous as the treatment of King Charles sounds to us today, I challenge the reader to imagine that our cures, our science, and our philosophy may one day be thought of as superstitious, like the pigeon dung of yesteryear. Only we don't know it yet. I can remember when cigarettes were considered good for the nerves. What will the new world scientist in 40 years think of radically killing living cells and tissues to cleanse the body of cancer? Examples abound. If the future is anything like the past, then just about everything will be proved wrong if we wait long enough.

Unfortunately, we are often completely unaware that our basic premises are beliefs, because we mistake them for facts, or **the way the world is**. I claim there is no such thing as a fact. Our Core Beliefs masquerade as fact. I call our Core Beliefs a framework, the thing that we hang all of our experience on. Some have called it a "World View." The philosopher/physicist Thomas Kuhn called it a "paradigm."[2] It's the *context* for our existence.

If we insist that thus and so is the way the world is, then we effectively cut ourselves off from thinking or perceiving anything other than what fits into our beliefs about reality. We will literally disregard any experience that doesn't work within our framework. I submit the following examples.

Have you ever wondered what happened to that gadget that you used to have in the junk drawer? You know you had one. You don't recall getting rid of it, but it seems to have vanished from your world. And what about the times you've opened the junk drawer and there is a gadget that you know you didn't have before, you've been meaning to buy one but you know you haven't yet, and there one is? You shrug your mental shoulders and move on. It got there somehow. Must be a logical explanation, even though you can't seem to get your mind around one. If it really bothers you, you will make one up, and then you'll believe your story. We are very proficient at editing what doesn't fit or what doesn't make sense to us.

Every person I have asked has admitted to having had the following experience: You have lost something important. Your keys, your driver's license, a favorite pocket knife or something like that. You look *everywhere.* You *must* have it. You look in all of the obvious places many times, and you look in many not-so-obvious places. You cannot find it. You finally give up because it is not to be found. You go do something else for several hours, maybe go to bed for the night. Next morning you wake up, you're combing your hair, you look down at the dresser and lo' and behold, there it is. Sitting right there in plain sight. In a place that you *know* you looked at least ten times.

So, what do you tell yourself? I have asked countless people this question and here is the answer that I always get: "I'm so stupid!" I then ask, "Why do you tell yourself that you are stupid?" And they say, "Because I didn't see it!" So I ask, "What are you assuming?" To which they look at me dumbfounded. They don't understand the question. The assumption is this: it was there when they looked *ten times* before and they didn't see it. Everybody knows this. **This is how it is.** If it is there now, then it **HAD** to have been there all along! How can this be so? How can you have looked **ten times** and not seen it if it was, in fact, there all along?

You might be thinking that someone else in the house probably moved your keys and then later put them on the dresser. This is the story you make up to explain your experience. What if, when queried, all other persons in your household deny having seen or touched your keys? What if you live alone? What then? You probably shrug your mental shoulders, assume

14

there must be some logical explanation and move on to more pressing concerns. But what is logical? Logical, by definition, is what complies with your system of what is and is not possible. If you accept that the keys were not there when you looked before and they are there now (your personal observation), without intervention from you or anyone else, then you do not have anything within your framework to hang this experience on.

I choose to believe that I am not stupid. The thing was not there before. I **know** that I looked there **ten times**. It was not there then. It is there now. I am fully aware that our collective belief system says that **it cannot be so**, yet that is my experience. That is also your experience. I am not willing to disregard my direct experience because it is not in vogue with current belief. I can't explain it, but I refuse to disregard it. There must be something wrong with the current belief. Perhaps there are multiple realities that we glide in and out of? Now you see it, now you don't. As long as you are good at editing what doesn't fit in with your framework, everything is a-okay.

Here's another little game that I have played with many people. Look at the figure on the next page and read what it says.

What does it say? Most people tell me that it says, "Paris in the spring." What it actually says is, "Paris in the the spring." Go look again. Did you edit out the second "the"? We all do this all the time. We have a need to have our experience make sense within our framework .

I often misread crossword puzzle clues. Once I have decided that I know what it says, I cannot see what is there and I misread it over and over again. It isn't until I have completed the rest of the puzzle and have the answer that I go back to the clue and realize that I had been misreading it all along. We literally do not see what is actually in front of us once we have decided that it is something else. We do this every day all day long. We automatically categorize everything that we see, hear, touch, or taste with reference to what we already know or believe or expect. Otherwise, the world can be frightening. Don't take my word for it, pay attention to your own experience.

If we habitually insert what we think is there, rather than seeing what is actually there, how can we know what really is "out there." Is there anything "out there"? Or are we making it all up in our minds? Could it be that there are a myriad of possible "out theres" to choose from and we only pick the ones we like (those that fit with our ideas)? Or are we just making it up as we go?

Have you ever looked at an object and have not been able to immediately make out what it is? It can be disturbing. You will stare and stare until you finally recognize the thing and can put it in proper context. Otherwise, it could be a monster.

Consider an account by Carlos Castaneda in one of his books about his adventures with the Yaqui shaman, don Juan. Part of the description of this book reads, "...don Juan, his mentor and friend, prepares him for the task of perceiving things as they are, instead of describing them by the words, conventions and standards of conventional, a priori ideas and language."[3]

Castaneda and don Juan were out in the desert and Castaneda saw something that he did not recognize. It was wailing and crying and making an awful fuss. It was gyrating and moaning and he was quite frightened. Then, finally, he resolved (forced?) it into a branch blowing around in the wind. He was quite proud of himself for figuring it out and he bragged to don Juan.

"I stared at it in complete and absolute horror. My mind refused to believe it. I was dumbfounded. I could not even articulate a word. Never in my whole existence had I witnessed anything of that nature.

Something inconceivable was there in front of my very eyes. I wanted don Juan to explain that incredible animal but I could only mumble to him. He was staring at me. I glanced at him and glanced at the animal, and then something in me arranged the world and I knew at once what the animal was. I walked over to it and picked it up. It was a large branch of a bush. It had been burnt, and possibly the wind had blown some burnt debris which got caught in the dry branch and thus gave the appearance of a large bulging round animal. The color of the burnt debris made it look light brown in contrast with the green vegetation.

"I laughed at my idiocy and excitedly explained to don Juan that the wind blowing through it had made it look like a live animal. I thought he would be pleased with the way I had resolved the mystery, but he turned around and began walking to the top of the hill. I followed him.

"...I began to talk about the branch, but he hushed me up. 'What you've done is no triumph,' he said. 'You've wasted a beautiful power, a power that blew life into that dry twig.' He said that a real triumph would have been for me to let go and follow the power until the world had ceased to exist.

"...He said that properly I should have sustained the sight of the live monster for a while longer. In a controlled fashion, without losing my mind or becoming deranged with excitation or fear, I should have striven to 'stop the world.'"[3] Apparently, seeing things as they really are, rather than stuffing them into our preconceived notions, takes some effort.

The framework of those of us who were born in the 20th century Western world is largely the result of science and the scientific revolution. There are some residual religious beliefs, but the primary concept that we have about **how the world is** comes to us from science.

What is the worst insult that anyone can give you? That you are an infidel? A pagan? No. The worst insult these days is for someone to tell you that you are irrational or unreasonable. We worship the rational and the reasonable. And the scientists own the market on rational and reasonable. Just ask them. Better yet, challenge them. At best, you will be called naive. At worst, you will be accused of wanting to plunge our society back into the superstitions of the dark ages. You either worship them and their version of reality, or you collapse the world into chaos and superstition.

Science is as much of a religion as any religion has ever been. The scientific priesthood bestows their thoughts and ideas upon the peasantry as if they were just bringing it down off the mountain. And woe to one who dares to question! Scientists themselves hesitate to openly question for fear of retribution [not getting their papers published or their proposals funded]. It's a conspiracy of fraud. I had to quit my job as a professional scientist in order to think. Some of the things science asks us to believe, with little or no proof, are more fantastic to a reasoning person than anything religion has put forward. And some of the really interesting data that scientists have uncovered is not believed by the scientists themselves.

1. Edward Dolnick, **The Clockwork Universe: Isaac Newton, the Royal Society, and the Birth of the Modern World**, Harper; 1st Edition, 2011.
2. Thomas S. Kuhn, **The Structure of Scientific Revolutions**, The University of Chicago Press, Chicago, IL, 1962.
3. Carlos Castaneda, **Journey to Ixtlan**, 1972, Washington Square Press; Reissue edition (February 1, 1991). --Note: Castaneda's work, as a "true" account of his anthropological experiences, has come into question. So what. It's still a fascinating story. And everything is a story.

2. A Bit of History
(Or the Gospel According to Descartes)

"The first precept was never to accept a thing as true until I knew it as such with out a single doubt."

--Rene Descartes, **Le Discours de la Methode,** 1637

So what is the framework for our beliefs? Most of our beliefs about how the world is are a result of ideas that were developed or proposed as theories to explain our experience in physical reality. In the Western world of the 21st century our primary world view is mechanistic, *i.e.* the universe is a machine.

In the Mechanistic world view there is no creator except, possibly, the creator of the machine itself. If there is a creator, then it created the universe like some giant clock at some time in the distant past. Since then the clock (universe) has been running on automatic. All observable phenomena must therefore be explained by causality or by random chance. I will call this the Clockwork Rule. It is forbidden to suggest that there exists anything outside of what we can observe or measure (God did it) when explaining our experience (data). This has evolved into: if our experiences cannot be explained by the laws of physics (based on causality and random chance), then they do not exist.

In order to understand the clock (universe), all we have to do is understand the parts and pieces, how they function, and how they fit together. This view is applied to the universe (physics and astronomy) and everything in the universe: the Earth (geology and archeology), the species (biology and anthropology), and even to our bodies (medicine) and minds (psychology and sociology). This is why we want to tear things apart to see how they worked, or to kill things to see what made them live.

How did this world view become accepted? It became accepted as a result of the scientific revolution. The scientific revolution was actually a gradual change in thinking over a period of a few centuries, perhaps starting with a publication by Nicolaus Copernicus in 1543 called *De Revolutionibus Orbium Coelestium* (On the Revolutions of the Heavenly Spheres) which posited a heliocentric (planets revolve around the sun) system instead of the geocentric (everything revolves around the Earth) system. This was a giant conceptual leap that affected our perceptions of the universe and our place in it. As an exercise, try to visualize how your beliefs and perceptions would differ if the geocentric model was currently accepted.

Because I shall be using the paradigm shift from the geocentric to the heliocentric models as an example later in this book, I shall describe the issues in some detail.

The Geocentric Universe & the Scholastic World View

During the Middle Ages the dominant Western world view was a combination of Christianity and Aristotelian (384-322 B.C.E) thought called the Scholastic world view. In this framework we (Earth) are at the center of the universe because, being made by God, we are the most important thing in the universe.

In the Scholastic world view, the Earth is made up of four elements, earth, fire, water and air. These elements each have their own proclivities. For example, things containing a lot of earth naturally fall to the center of the universe (Earth) while fire makes things rise. Objects in the heavens (sun, stars and planets) were made of a fifth, divine element with the circle as its natural motion. This was a very different way of thinking about how the world is and our place in it.

The Greeks were keen observers and they were also experts in geometry. They considered geometry sacred, particularly the circle which has no beginning and no end. The Greeks observed that the sun, stars and planets all appear to move in the sky from east to west. From this they deduced that the sun and all of the planets and stars revolve around us in what they assumed were sacred, circular orbits. They noticed that the motion of the stars was different from that of the sun, which was also different from that of the planets.

The Greeks postulated that the stars were fixed on a great sphere and this sphere rotated on an axis with the Earth at its center. The sun occupied a different sphere and the planets each on their own sphere. The planets were thought to be stars, but because their motion was not consistent with that of the other stars (in fact they even sometimes appeared to move backwards in the sky) they were called "the wanderers." The Greek word "planet" literally means "wanderer." The Greeks concluded that the Earth was spherical by observing the shape of the shadow cast by the Earth onto the moon.

The Greek philosopher Aristarchus (310-230 B.C.E.) proposed the first known heliocentric model of the solar system around 280 B.C.E. He deduced this by using geometry to estimate the relative distances between the sun, Earth and moon. From these and the angular diameter of the sun and moon, he calculated the relative sizes of the sun and moon. While his numbers were not in accordance with current values, he found that the sun was much bigger than the moon and the Earth. This lead him to postulate that the sun and not the Earth was at the center of the universe because, being so much bigger, it must be more important. Because of this, "the Stoic philosopher Cleanthes (c. 301-232 B.C.E.) declared that it was the duty of Greeks to indict Aristarchus on the charge of impiety..."[1] So it goes.

Besides being heretical, Aristarchus' heliocentric model was not accepted on scientific grounds. First, if the Earth rotated around the sun then the position of the stars would change when viewed from different positions in the orbit (known as parallax). This was not observed because the stars are so distant that the parallax is too small to observe without telescopes, but this was not known at that time. Second, if the Earth is rotating about its axis from west to east to account for day and night, then there should be a great wind always in the direction opposite the Earth's motion, east to west. This is obviously not observed. Finally there was the "tower problem." If an object were dropped from a very high tower and the Earth is moving, then the object should not fall at the foot of the tower but some distance from it to account for the motion of the Earth. This was a conceptual error that was not resolved until Einstein pointed out that all motion is relative and there is no way to tell whether your whole environment is moving so long as everything in it is moving at the same velocity. Thus, the geocentric model was believed for a couple thousand years or more.

Around 150 C.E. in Alexandria, Egypt the astronomer Claudius Ptolemy developed a detailed model of the geocentric system that (at the time) accurately predicted the positions of the planets and stars. In his model, Ptolemy dispensed with the Greek idea of the sun and the planets being fixed on rotating spheres. He did this because he was forced to invent an extra motion for the planets to account for the observed retrograde (backwards) motion of the outer planets and the change in apparent brightness of the planets.

25

In the Ptolemaic model the planets not only rotated about Earth in a circular orbit, they also moved in smaller circles (called epicycles) within that orbit. Ptolemy's model persisted for some 1300 years, but there was one major problem. While it accurately predicted the positions of the planets, after a hundred years or so, those predictions were a little off. After a few hundred years they were off by quite a bit. The solution was to simply recalibrate (change the numbers to the current values) and march on.

The Heliocentric Universe & the Mechanistic World View

Over one thousand years later, the Polish astronomer Mikolaj Koppernigk (known as Nicolaus Copernicus 1473–1543 C.E.) found the Ptolemaic model very unsatisfactory because the epicycles made it cumbersome and the predictions were inaccurate over a long time period. Copernicus resurrected the heliocentric model of Aristarchus, which he found much more aesthetically pleasing. Copernicus suggested that we don't experience the big wind from the Earth's spin because the Earth's atmosphere spins along with it. Because telescopes had not yet been invented, there was still the problem of the unobserved parallax. In addition, there was no way of knowing which model was more accurate since it took a few hundred years to notice the inaccuracies.

When two theories exist that explain the same phenomena, the general rule is to choose the simplest one. This is known as Ockham's Razor, after the monk-philosopher William of Ockham (c. 1287–1347). By this rule the Copernican model should have been immediately adopted. This did not happen because the Holy Roman Church (the major superpower of the time) was against it. Copernicus himself, being very devout, was bothered by it but he argued that it is natural and logical to have the sun at the center of the universe because the sun is the source of light and life.

Proponents of the heliocentric model were not smiled upon by the Church. In the 1600's both Galileo Galilei and Giordano Bruno were tried by inquisition and found guilty of heresy for promoting the heliocentric model of the solar system. Bruno was burned at the stake while Galilei recanted and was placed under house arrest until his death. Thus was the fate of those who disagreed with the priesthood in those days. The Vatican formally pardoned Galilei in 1992. It was not until 1822 that the Church of Rome accepted the Copernican theory of a heliocentric solar system.

Rene Descartes (1596-1650) was probably the main historical proponent of the mechanistic world view. Descartes and other philosophers who followed in his path helped to establish a mechanistic paradigm for understanding natural phenomenon. This approach stood in marked contrast to the philosophy of the time, which at least implicitly involved the existence of

27

spiritual or otherwise non-physical forces playing a significant role in natural processes.

The advent of the mechanistic world view, as proposed by Descartes in the first half of 1600's and furthered by Isaac Newton (1642 -1727) in the second half of 1600's and early 1700's, completely changed how science was conducted. Up to this time, science and natural philosophy were considered one and the same. Subsequently, empiricism became the primary focus of science, requiring more complex equipment and laboratories. This had the eventual result of moving scientific inquiry from the parlors of the idle rich to university laboratories.

The mechanistic world view was in direct opposition to vitalism, a popular philosophy of the day that was championed by Lady Anne Conway (1631-1679), Henry More (1614-1687), Francis Mercury van Helmont (1614–1698) and Gottifried Wilhelm von Leibniz (1646-1716). According to vitalism the universe is made up of tiny, indivisible particles called 'monads,' each possessing a life force. This life force exists in all organisms and is caused and sustained by a vital principle distinct from all physical and chemical forces and is self-determining and self-evolving.

The validity of these two doctrines was hotly debated in the parlors of the intelligentsia during the 1600's. In 1687 Isaac Newton published *Philosophiæ Naturalis Principia Mathematica* (Mathematical Principles of Natural Philosophy). As a result of careful observation, Newton was able to explain many previously unexplained phenomena. In *Principia*, he basically invented calculus and the forces of gravity and inertia (exemplified in the collision of two masses). This was a huge accomplishment at that time and was instrumental in swaying the debate in favor of mechanism and the scientific method. Thus is our legacy.

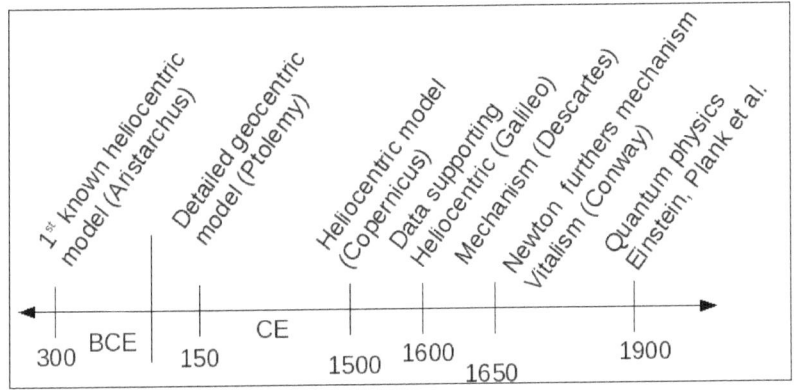

Figure 2.1: A brief time-line of scientific ideas.

1. Jean-Louis & Monique Tassoul, **A Concise History of Solar and Stellar Physics**, Princeton University Press, Princeton, NJ, 2004.

3. Philosophy and Science
(Or The Gospel According to Popper)

"Science is all about truth. You gather your evidence and
logically prove your claims."
--Jerry McNerney, PhD (Mathematics), on the difference between
science and politics, interview in Wired Magazine, Issue 15.03,
March 2007.

Because science and philosophy have historically
been intertwined, we need to take a brief detour into
philosophy. It should be clear from the previous
chapter that science was developed as a discipline in an
attempt to answer some basic philosophical questions.
These are: How do we know that there is anything "out
there" independent from our subjective experience?
And What is real (or true)? The astute reader may have
noticed that other disciplines have also set about
answering these questions. They are called religion.
The difference supposedly being that religion relies on
belief while science and philosophy rely on logic.

Philosophers have debated for millennia about
what we can and can't know and about what is and isn't
real. Plato, Socrates and Pythagoras claimed that the
only true reality is idea. Physical reality is a
manifestation of our ideas. Whatever we experience is,

by definition, less than our ideas (idealism). More recently, philosophers have debated about what can or cannot be proved. How can we justify using past observations as a basis for generalizations about what we have not yet observed. It has been acknowledged by most[1] philosophers that nothing can be proved true. One illustration of this can be seen in what is known as *The Ravens Problem.*

The philosophy game is played by making a statement (hypothesis) then using the rules of logic to prove or disprove the statement. The ravens problem goes like this:

Statement# 1: All ravens are black.

How do we go about proving our statement is true? We employ the scientific method of observation. We go out in the world and look for ravens. We need find only one non-black raven to prove the statement false, but how many black ravens need we find to prove it true? No matter how many black ravens we tally, we cannot guarantee without doubt that the next raven we see will be black. We can, at best, assign a probability that the next raven we see will be black, but we cannot prove the statement is true by observing black ravens.

You might say, "Come now, don't be ridiculous. Of course there is some small probability that the next raven we see will not be black, but we have observed so many black ravens we can be confident that we have proved our statement." Okay then. By strict rules of logic, the statement

Statement #2: All non-black things are not ravens

is equivalent to the previous statement, all ravens are black. Therefore, if finding black ravens "proves" statement #1, then finding green tennis balls "proves" statement #2. Because statement #2 is equivalent to statement #1, I can "prove" that "all ravens are black" by finding green tennis balls. Or white shoes. I don't even have to go outside and find ravens to prove my hypothesis. Abandon hope, all ye who worship logic!

The Ravens Problem highlights the weaknesses of inductive logic. Inductive logic is when you have some data (observation of black ravens) from which you form an hypothesis that explains the data (all ravens are black). A large number of possible resolutions to the ravens problem have been proposed. The various arguments for and against each will make your head spin, but I don't believe there has been consensus among philosophers. And anyway, other paradoxes concerning inductive reasoning exist, along with more gnarly discussions. The bottom line is that it seems that we cannot confirm or prove a statement (hypothesis) made by inductive reasoning.

The situation is not improved by switching to deductive logic. Deductive logic starts with a premise or premises that are *assumed* true. Given that the premises are true, we then determine what can logically follow and call these conclusions. The premises themselves are unproven and unprovable, they must be accepted at face value, or by *faith*, or for the purpose of exploration. For example, using deductive reasoning we simply state without proof that:

1. All ravens are black.

We have a friend who has a pet raven called Berthe. We can logically deduce that Berthe is black, without ever having observed Berthe. We have no way of knowing whether or not the premise is true, but assuming that it is, Berthe is most certainly black. Notice that the reverse is not true, *i.e.* just because Berthe is black does not mean she is a raven. All scientific theories are built up in one or the other of these two ways.

One of the jobs of the philosophers is to establish the rules for what does and does not constitute an acceptable scientific theory. While the philosophers are by no means in agreement, Karl Popper's (1902 – 1994) philosophy is most widely accepted. Among scientists, that is. Popper's view of science is highly idealistic, which is probably why it appeals to scientists. In fact, I would venture to guess that most scientists don't know that there *are* any other philosophies concerning science! They've probably never heard of Popper either.

Popper believed that induction has no place in science. According to Popper, scientific theories should be formed by deduction. He agreed that we cannot confirm or prove a theory true, but that a good scientific theory has the possibility of being falsified or disproved. The experimental data must trump the theory, no matter how beautiful the theory. Popper had an ax to grind with socialism, which he said cannot be falsified because no matter what happens, a Marxist can somehow fit it into her theory. Popper referred to such theories as pseudo-science. Any theories that cannot be subjected to falsification tests do not qualify as scientific, in Popper's view. For scientific theories, we set up experiment after experiment to try to prove them false. If we fail to falsify the theory, then it is not necessarily true, but it is robust and can be relied upon to give accurate predictions. The theory then becomes a working model.

If any one scientist is too egotistical and too attached to his or her particular conjecture to reject it based on experimental data, that should pose no problem as presumably there are other scientists who are wedded to different conjectures, so that it should all work out fine in the end. Even if everyone refuses to believe the new data, a new theory will eventually emerge when the old crowd dies off. "For Popper, a good or great scientist is someone who combines two features... The first feature is an ability to come up with imaginative, creative, and risky ideas. The second is a hard-headed willingness to subject these imaginative ideas to rigorous critical testing. A good scientist has a creative, almost artistic, streak and a tough-minded, no-nonsense streak. Imagine a hard-headed cowboy out on the range, with a Stradivarius violin in his saddlebags. Perhaps ... you can see ... the reasons for Popper's popularity among scientists."[2]

The criteria for a good scientific theory therefore, is that it has held up to intense scrutiny and has not yet been falsified. This does not make it "true" or "real" but it qualifies as a good map of reality. It is unfortunate that many, scientists included, mistake the map for the territory. And it is shocking that many, scientists especially, make the claim that thus and so has been confirmed or proven.

A criticism of Popper's philosophy of science is what I will call Popper's Decision Problem. It goes like this: Suppose you want to build a rocket. There exists a model for thrust and lift and so forth that has been used successfully in the past to build rockets. But a new model has come along that has never been tested. According to Popper, there is no reason to believe one model is more true than the other, since confirmation is not possible. Neither model has been shown to be false. Which model would you use to build your rocket? Popper was aware of this problem and was unable to give a response that did not involve some sort of confirmation.

The criteria that scientists *claim* they use for a good scientific model is very close to Popper's ideal.[3] They are as follows:

1. The model must agree with observed data.
2. The model must make predictions that allow it to be tested. It must be possible to disprove the model. The model may need to be modified to fit new data or it may need to be discarded entirely.
3. The model should be aesthetically pleasing. It must be simple, neat and contain the fewest possible assumptions.

Criterion #3 is not strictly necessary, but is used in the same manner as Ockham's Razor, *i.e.* when two competing theories explain the same data, choose the most aesthetically pleasing, which is also the simplest. It is supposedly also used as a bellwether, *i.e.* if a theory gets too cumbersome with many *ad hoc* adjustments and modifications then it is ugly and needs to be replaced. Such was the case with the Ptolemaic geocentric model of the solar system with its epicycles and recalibrations.

I claim that Popper's view of how science is done is idealistic for the simple reason that, while this may be how science *should* be done, it is not. Suppose our statement, All Ravens Are Black is given as true. It follows that if I go outside and see a raven, it will be black. If Popper were correct, then a scientist trying to falsify our statement would go outside and actively look for white ravens. Or green and purple ones. What *in fact* would happen is that our scientist would go outside and look for black ravens *in support* of the statement.

Suppose he is out there looking for black ravens and a white one flies by. He likely wouldn't see it because he is only looking at *black* birds and then determining if they are ravens. He is paying no attention whatsoever to white birds. If he does happen to see one, he will wonder if he made a mistake. The sun was in his eyes and it only *seemed* white from that angle, so it must've been black after all. Thus he will justify discarding that data point and continue his search for black ravens.

Suppose further that our intrepid scientist runs across yet another white raven. Now he is in a pickle because all the other scientists in the world also believe that ravens are black. If he publishes his results they will not believe him. He will either abandon the project and go do something else or he will go ahead and try to publish. The reviewers will recommend that his paper not be published because everyone knows that ravens are black. They will question his methods or impugn his integrity. Was he wearing sunglasses or another optical device when the so-called observation was made? Did he record the time, date, temperature, atmospheric pressure and GPS coordinates so other scientists can repeat his observations? Scientists who have a stake in the theory (a lifetime of work and publications) will claim that he misunderstood or misinterpreted his observations. The raven probably had some snow on it or something. Kill the messenger.

Suppose still further that a second scientist, being more open-minded than her fellows, went out where our first scientist was and she also spotted a white raven. Now we have a newspaper headline: Scientists Find White Raven; Black Raven Theory In Trouble. Then the public never hears about it again, but six months later the Raven Theory is still intact. It now looks like this:

Theory: All Ravens Are Black

Codicil: On the first Sunday after the first Full Moon after the Spring Equinox, white ravens might possibly be observed.

At first glance, it appears that scientific theories cannot be falsified. We have already established that they cannot be confirmed. What a deal.

Another reason that scientists agree with Popper's philosophy of science is that it gives them an excuse to reject anything they don't like or don't understand. Sort of like Popper and socialism. Just say it isn't science and move on. A case in point is Rupert Sheldrake (b. 1942).

Rupert Sheldrake is a renegade biologist. He started out as a perfectly respectable biologist until he got obsessed with all the things we don't understand that we pretend to understand. Or things we pretend don't exist or just plain don't try to understand. Such as: how do pigeons home; how do pets know when their owners are coming home; how can we tell when we are being stared at and so on. He has written several books, one of which is called **A New Science of Life**, where he proposes a "morphogenetic field" containing information that can be accessed via "morphic resonance."

In Popper's world scientists would grab hold of this theory, figure out what sort of predictions can be made and go test them (by falsification, of course). What actually happened was that the late Sir John Maddox wrote in an editorial in *Nature* (1981), "This infuriating tract... is the best candidate for burning there has been for many years." In an interview broadcast on BBC television in 1994, Maddox said, "Sheldrake is putting forward magic instead of science, and that can be condemned in exactly the language that the Pope used to condemn Galileo, and for the same reason. It is heresy." There's a fine example of open-minded scientific inquiry.

Who among us has not experienced the sense of being stared at? Yet scientists will tell you it does not exist. For many years scientists claimed that asteroids do not exist because rocks do not fall from the sky. Suppose that a rock fell from the sky into your yard. You pick it up and take it to the local university where the scientific priests can be found and they tell you that it could not have fallen from the sky. Maybe a bird dropped it. Do you believe them? Or do you believe your own direct experience? If you believe them, you begin to doubt that you saw what you saw. And if any scientist dares to explore rocks falling from the sky, she runs the risk of being publicly humiliated, denied funding, and encountering difficulty publishing. She'd better have another source of income. What part of this do you consider logical, reasonable, or rational?

There are other philosophers who expound different ideas about how science is or should be done. Because in these days philosophers are not scientists, they don't actually know how science is done. And scientists don't listen to philosophers anyway. With one notable exception.

In 1962 Thomas Kuhn (1922 – 1996) published a book on the philosophy of science called **The Structure of Scientific Revolutions**. Kuhn started out as a physicist, then he wandered into the history of science and from there into the philosophy of science. Kuhn claimed, as I do, that our world view is a result of the current scientific paradigm. He actually coined the term "paradigm" in this sense. The paradigm is the big picture, which may include many theories and hypotheses. Classical Newtonian physics is an example of a paradigm that includes kinematics (baseballs and such), optics, electrodynamics, gravity and the like. Each of these has various theoretical models that yield predictions and so forth.

According to Kuhn most all of science is practiced as "normal" or "paradigm-based research," *i.e.* research within the currently accepted scientific paradigm. No-one actively tries to falsify or overthrow the paradigm, nor should they according to Kuhn. One needs to have a paradigm as a starting place within which to operate. It's our belief system. "No part of the aim of normal science is to call forth new sorts of phenomena; indeed those that will not fit the box are often not seen at all. Nor do scientists normally aim to invent new theories, and they are often intolerant of those invented by others. Instead normal-scientific research ... seems an attempt to force nature into the preformed and relatively inflexible box that the paradigm supplies." [4]

"Normal science" consists of three major experimental and observational endeavors, all of which constitute fact gathering in support of the current paradigm. They are:

1. Determining physical properties with greater precision (stellar positions, specific gravity of materials, compressibility of materials, spectral properties, electrical conductivity of materials, etc.).

2. Demonstrating agreement between theoretical predictions and experimental data to greater accuracy by improving experimental apparatus or finding new ways to demonstrate agreement.

3. Articulation of theory:

a. determining physical constants with greater precision (speed of light, universal gravitational constant, Coulomb's constant etc.)

b. exploring empirical laws (Boyle's law, *e.g.*)

c. determining the limits of applicability of laws within the paradigm.

Normal science also consists of three corresponding theoretical pursuits:

1. Using the existing theory to calculate expected values of physical properties that can then be measured.

2. Adjusting the theory to more closely agree with experimental observations or modifying the theory for special cases (adding air resistance to kinematic calculations, or modifying the value of a constant in the calculations *e.g.*).

3. Reformulating the theory to extend the model (adding another mass to calculate the gravitational interaction for three masses instead of two, *e.g.*).

Two things are immediately noticeable about normal science. One is that much of it is excruciatingly boring. Even scientists find it boring. This is why they invented positions now filled by graduate students and post-docs. The other, more important thing is that no-one is trying to disprove anything. All of the effort is going into supporting the current paradigm. Popper's Decision Problem is moot because there are no theories that do not support the current paradigm.

How then do theories and paradigms change? In the course of normal science small puzzles pop up; little pieces of data that do not fit within the paradigm. Often these are simply not seen. Kuhn compares this with a psychological experiment published in 1949 by J.S. Bruner (b. 1915) and Leo Postman (1918-2004).[5] In this experiment subjects were asked to look at a sequence of short exposures of playing cards and identify them. Inserted into the deck of playing cards were anomalous cards, a red six of spades and a black four of hearts. Responses to the anomalous cards were divided into four categories: they were reported as normal (*i.e.* the red six of spades was simply reported as a six of spades—the authors called this a "perceptual denial."[6]); they were reported as anomalous but incorrectly (*e.g.* a purple four of hearts); the subject knew there was a problem but couldn't figure it out (one subject said, "I don't know what the hell it is now, not even for sure whether it's a playing card."); and finally some subjects reported what they saw (a red six of spades) though they had to be exposed to it many times before they saw it correctly. Kuhn argues that scientists, being human after all, do the same.

Once the anomalies are seen and identified, scientists work to improve the detection apparatus. If that doesn't resolve the puzzle, they modify the existing theory to accommodate the data for the special case. Usually the great discoveries of science come from solving these little puzzles. Occasionally an anomaly arises that resists all efforts to resolve it within the paradigm. Every so often these resistant anomalies build up into a critical mass where scientists start to lose faith in the paradigm. Kuhn calls this a crisis.

Only during a crisis are new theories formed outside of the working paradigm. Once a new theory arises that can satisfactorily predict the anomalies, along with the data that were not in question, then a "paradigm shift" occurs that changes our world and our world view. This was the case in shifting from the geocentric view to the heliocentric view and in the revolution of quantum mechanics vs. Newtonian mechanics.

According to Kuhn it is a good thing that scientists are resistant to change because our knowledge is added to considerably during the course of normal science and it wouldn't do to distract the scientists too much from their belief system. We can't be having paradigm shifts every time our data doesn't come up to snuff. And anyway everything works out in the end because science is a self-correcting system that will root out its problems even if entire generations must die off before new ideas can be accepted.

This would be fine if you believed that our knowledge is always increasing and we are always getting closer to Truth than we were before. People assume this is so; that we are standing at some sort of pinnacle of knowledge and all who have gone before us are poor, ignorant fools. They grab the moral high ground, bludgeoning us with their version of reason and logic and insinuate that our **only** other choice is superstition, weeping and gnashing of teeth. They accuse anyone who dares to question science of trying to throw us all back into the dark ages.

Not so, says Paul Feyerabend (1924-1994). First of all Feyerabend saw science as a creative endeavor. The great scientists were anarchists, they were unafraid of breaking rules laid down by philosophers and they used any and all means available to them to aid in scientific discovery. In Feyerabend's ideal world anything goes and any attempt to force scientists into a prescribed method will only serve to dampen creativity and create stultifying science.[7] But, while science started out to free us from the thought police in the One True Religion (Christianity), it has itself turned into the oppressor. He argues that we now need to be freed from the grip of the scientific establishment and he advocates a separation of science and state, like any good democracy.[8]

Feyerabend claims that there are no objective reasons for preferring science to other traditions and ways of knowing. He says that, "Most intellectuals have not the foggiest idea about the positive achievements of life outside Western civilization. What we [have] in this area are rumors about the excellence of science and the dismal quality of everything else. ...Western science has now infected the whole world like a contagious disease... Western civilization was either imposed by force, not because of arguments showing its intrinsic truthfulness, or accepted because it produced better weapons. ... [R]ationalists have devised [arguments] to overcome difficulties. For example, they distinguish between basic science and its applications: if any

46

destroying was done, then this was the work of the appliers [the politicians], not of the good and innocent theoreticians. But the theoreticians are not that innocent. *They* [emphasis his] are recommending analysis over and above understanding, and this even in domains dealing with human beings; *they* [emphasis his] extol the 'rationality' and 'objectivity' of science without realizing that a procedure whose main aim is to get rid of all human elements is bound to lead to inhuman actions. Or they distinguish between the good which science can do 'in principle' and the bad things it actually does. That can hardly give us comfort. All *religions* [emphasis mine] are good 'in principle' – but unfortunately this abstract Good has only rarely prevented their practitioners from behaving like bastards."[9] Feyerabend dismisses arguments in defense of science such as 'science knows best' (feeble) and 'science works' because, "Science works sometimes, it often fails and many success stories are rumors, not facts."[10]

Scientists of today are largely ignorant of what the philosophers think or say. Feyerabend noticed this and he called us "uncivilized" because of it. I think that Popper's version of science is unrealistic but it's how scientists see themselves, although they don't seem to understand the full implications (nothing can be known for sure). Kuhn's version of science is how science is actually done, whether it should be (as Kuhn argues) or not. I agree with Kuhn that theories should not be

47

thrown out at the first sign of trouble because they are still useful as models (and they are all only models) as we shall see. Feyerabend's version of how science should be done seems the most fun and promising, not to mention open-minded. I leave this chapter with Godel's Theorem (Kurt Godel, 1906-1978).

Godel's Incompleteness Theorem is a mathematical theorem that he did not intend to be applied outside of mathematics. However many have argued that his theorem is applicable to any system of logic, which includes physics. Godel's Theorem states that we cannot prove the veracity of a system from within the system. Another way of saying that is: We cannot prove that the system itself is true using the axioms of the system. Now, in plain English: We made up the rules but we cannot use the rules to prove the truth of the rules.

For example, algebra is a system with made-up axioms, namely the commutative and associative properties of addition and multiplication of numbers (numbers, themselves are made up!). This system of manipulating numbers is internally consistent, so long as we follow the rules. However, we cannot prove the truth (or usefulness in the physical world) of the system of algebra using the axioms of algebra. We cannot even prove that algebra is the only useful way of manipulating numbers in the world (uniqueness) using the axioms of algebra. If, using the axioms, we were to encounter an inconsistency, then we should begin to suspect the veracity of the system, or at least the axioms. Thus we can only show that a system is false from within the parameters of the system.

Once again in plain English: as long as we are in physical reality, there is nowhere to stand outside of physical reality to determine the truth of our ideas about physical reality. We cannot know that anything is True. We can't even know that there is a physical reality "out there" independent from our thoughts and ideas about it. So, let us not get carried away with our ideas and take our models too seriously. They are road maps, nothing more. The terrain may be entirely different when viewed from the ground. Direct experience is the only thing we can know for sure, and even that can be rationalized away.

1. Peter Godfrey-Smith, **Theory and Reality**, University of Chicago Press, Chicago, IL, 2003.
2. Ibid., p. 62.
3. See, for example: Karl F. Kuhn and Theo Koupelis, **In Quest of the Universe**, 3rd edition, Jones and Bartlett Publishers, Sudbury, MA, 2001, p.37.
4. Thomas Kuhn, **The Structure of Scientific Revolutions**, 2nd edition, the University of Chicago Press, Chicago, 1970, chapter 3.
5. Kuhn, Ibid., chapter 6.
6. J.S. Bruner and Leo Postman, "On the Perception of Incongruity: A Paradigm," *Journal of Personality*, XVIII, 1949, 206-223.
7. Paul Feyerabend, **Against Method**, 3rd edition, Verso, London, 1993 (first published in 1975 by New Left Books).
8. Ibid., appendix 2.
9. Paul Feyerabend, **Farewell to Reason**, Verso, London, 1993, chapter 12 section 4.
10. Ibid., chapter 12 section 3.

Part II

The Trouble With Physics

4. Space and Time
(Or The Gospel According to Newton & Einstein)

*"If nobody asks me, I know what time is, but if I am asked then
I am at a a loss what to say."*
--St. Augustine, **Confessions of St. Augustine,** Image Books
Edition, NY, Bk. 11, Ch. 14, 1960

Physical science, or physics, is the study of how the physical world works. These days physics consists of a medley of different theories. An embarrassment of theories. These theories don't compete with one another, rather each one is only applicable to certain, restricted situations. If the situation changes, we must go and grab a different theory off the shelf. It's enough to make a person crazy.

In an effort to remedy this situation, there has been an ongoing search for One Theory (a Grand Unified Theory) that describes the workings of the entire known universe. This theory must reduce to all of the currently known theories when the appropriate restrictions are applied for the situation. We impose this requirement because the known theories work. We use them to build faster computers and other useful gadgets. Thus far the effort to find the One Theory has failed utterly. This should be a clue that we are entirely

53

on the wrong track. Recall from Chapter 3 the criteria for a good scientific theory:

1. It must agree with observations.
2. It must make accurate predictions that are testable.
3. It must be aesthetically pleasing.

Taken individually, the current theories all satisfy criteria #1 & #2, and at least some satisfy #3, but taken as a whole, the situation is a mess.

As previously mentioned in Chapter 3, criterion #3 is not strictly necessary. However, history has shown that when theories become messy, with many *ad hoc* corrections and restrictions, there has been a major conceptual error made somewhere. Whenever we have tried to force reality to fit our theories, we have had to radically change our thinking about reality, along with our theories. The heliocentric model of the solar system and the quantum theory are cases in point. I believe we are at another such juncture now.

To unravel where we might have gone wrong, it is necessary to begin at the beginning of physics as we know it, with Sir Isaac Newton (1642 -1727). Newton wanted to understand the principles behind natural motions and interactions of physical objects. The motion of an object can only be defined as a change in its position in relation to something else, but relative to what?

Do we define the motion of an object as change in its position relative to all of the other objects in the universe? (This is now known as Mach's principle, after Ernst Mach, 1838 – 1916.) Or can an object's motion be defined independently from other objects, relative to empty space? What is empty space? Is space a fixed container that we (or God) put things into? Or do only the objects exist and we invent space to separate and define them?

Newton decided that space exists in its own right and there is some frame of reference that is perfectly still that he called Absolute Space. All motion is defined relative to Absolute Space. Newton worried a little that he could not find, measure, or even define this Absolute Space, nevertheless he reasoned that it must exist as the backdrop for objects to occupy and for events to happen. Furthermore, Newton's Absolute Space is flat (Euclidean), meaning that the shortest distance between two points is a straight line. No curves or other funny stuff are allowed in Absolute Space.

Whether we define the motions of an object relative to other objects or to the non-object we call space, we need to invent another artifact called time. How can anything change its position without time? Newton defined the motion of an object as a change in its position in space in one instant to a new position in space at a later instant.

Time is measured by some repeating (periodic) pattern that can be replicated, such as a pendulum or an atomic transition. This periodic interval is then converted to seconds (or some multiple thereof) and applied to every event. Newton assumed that the interval was the same for everything in the universe regardless of their respective positions or velocities or whatever. This he called Absolute Time. If all of this seems trivial and obvious to you then you haven't thought about it enough, but fear not, it will rear its ugly head over and over again in this book.

Having defined the parameters of Absolute Space and Time, Newton then had to decide on the most fundamental quantity of Nature. He chose force. Likely he thought this was the obvious choice based on observations; *i.e.* if you kick a ball it moves and if you don't, it doesn't. Newton then formulated his three famous laws.

Newton's first law: An object in motion continues in motion forever unless acted upon by an external force. This is also known as the Law of Inertia because an object at rest will stay at rest unless acted upon by a force. Objects don't move or speed up or slow down or change direction on their own.

Newton's second law: The acceleration, a (change in motion) experienced by an object is directly proportional to the force, F applied and inversely proportional to its mass m, or $a = F/m$. (note: this is more commonly written $F = ma$, bold indicates vector quantities)

Newton's third law: For every action (force) there is an equal and opposite reaction (force).

This is an example of a deductive logic system. Newtons laws are stated without proof and *assumed* true. They are not derivable or provable. All of what is now known as classical or Newtonian physics follows from these premises. Classical physics makes up the bulk of a physics undergraduate education.

The most widely applied of Newton's laws is #2, which is a statement of how the position of an object will change when acted upon by an external force. The acceleration given to the object will depend on its mass. The mass is a measure of its *resistance* to move, called inertia. The larger the mass, the harder it is to get it to move. For this reason, the mass in Newton's #2 is called the *inertial mass*.

What is mass really? We don't actually know. Its the stuff that makes up the world we experience, but when we look at smaller and smaller pieces of it, it disappears entirely. Elementary particles are little bits of nothing that defy description (more on this in Chapter 6).

We *define* mass as the density per volume of some stuff. More stuff in the same volume results in more mass (*i.e.* a cubic centimeter of lead has more mass than a cubic centimeter of pudding because the lead has more stuff (higher density) crammed into the same amount of space. It turns out that not only can we not describe the motion of an object without inventing space, we can't define the object itself without it.

After formulating his laws and inventing calculus to describe the motion of objects, Newton started wondering about the moon. What holds it up there? Why doesn't it fall and crash into the Earth like everything else that gets thrown up there? What causes things to come crashing back to Earth anyway? According to Newton's #2, there must be a force in there somewhere, else balls thrown up in the air would continue on their way forever and ever (by Newton's #1). Where is this force? Nothing touches the ball (or the moon) after it leaves our hand.

Newton reasoned that, according to his #1 law, the moon should continue in a straight line off into the universe (having been somehow set in motion in the first place). Because it doesn't, then according to Newton's #2, there is a force holding it in orbit around the Earth. In a stroke of genius, he invented one. He called it *gravity*. Newton eventually came up with an equation for this force that we now call Newton's Universal Law of Gravity. The magnitude of the gravitational force, F between two masses (m_1 *and* m_2) is equal to the product of the two masses and divided by the distance between them, r squared,

$$F = \frac{G m_1 m_2}{r^2}. \tag{4.1}$$

This equation is not derived from Newton's laws, he just made it up. The capital G is a fudge factor called the Universal Gravitational Constant. I won't bother you with its value because, if we found some error in any of the other quantities, we would just jiggle the value of G to make the numbers work. As you will see if you keep reading, we do this all the time.

According to Newton's #3, the force on mass m_1 due to m_2 is equal and opposite to the force on m_2 due to the presence of m_1. To determine the force of gravity between the Earth and the moon, we simply plug in the values for the masses of the Earth and moon (how we know these is another story) and the distance between. To determine the force on a ball that pulls it back to Earth after we have thrown it we use,

$$F = \frac{G m_1 M_E}{R_E^2}. \tag{4.2}$$

where m_1 is the mass of the ball, M_E is the mass of the Earth and R_E is the radius of the Earth (see box 4.1). If we define

$$g \equiv \frac{G M_E}{R_E^2} \tag{4.3}$$

as the acceleration due to the force of Earth's gravity at or near the surface, then $\boldsymbol{F_G} = m_1 \boldsymbol{g}$. But this equation looks like Newton's #2, $F = ma$ with $a = g$. The gravitational mass is equivalent to the inertial mass! Furthermore, mass has some magical property of imposing a force on other masses over a distance without ever coming into contact with them. Even Newton found this spooky.

Box 4.1 Mathematical Tricks of the Trade

In deriving Eq. (4.2) two tricks were employed. We really should divide the Earth up into a gazillion ball-sized pieces and calculate the force on the ball due to each piece of the Earth and then add up all the forces. But the math is too hard, so we invoke the "rigid body" rule. As long as the individual pieces of the Earth aren't moving around within the body itself (we ignore the oceans), we can approximate the forces and the motion of a rigid body by replacing it with a point mass placed at the center (center of mass) of the rigid body. In effect, we assume that all of the Earth's mass is concentrated at the center of a sphere of radius R_E! Physics is rife with these sleight-of-hand calculations. Occasionally the mathematicians howl, but the physicists don't bat an eye.

The second trick was to replace r in the denominator with the Earth's radius, R_E. Strictly speaking, it should be the distance between the Earth and the ball, $R_E + h$, where h is the height of the ball above the surface of the Earth. But we assume that $h \ll R_E$ therefore $r \sim R_E$.

Albert Einstein (1879 – 1955) found it spooky too. Because the shortest distance between two points on a sphere is a curve, Einstein reasoned that the orbital motion of the moon about the Earth (or any orbital motion) can be explained by imagining that space is curved. This is rather difficult to imagine in three dimensions, but in two dimensions imagine putting a bowling ball in the middle of a waterbed. The surface of the waterbed mattress will curve toward the center of the bowling ball from all directions. Next imagine tossing a ping-pong ball at a right angle as shown in Figure 4.1.

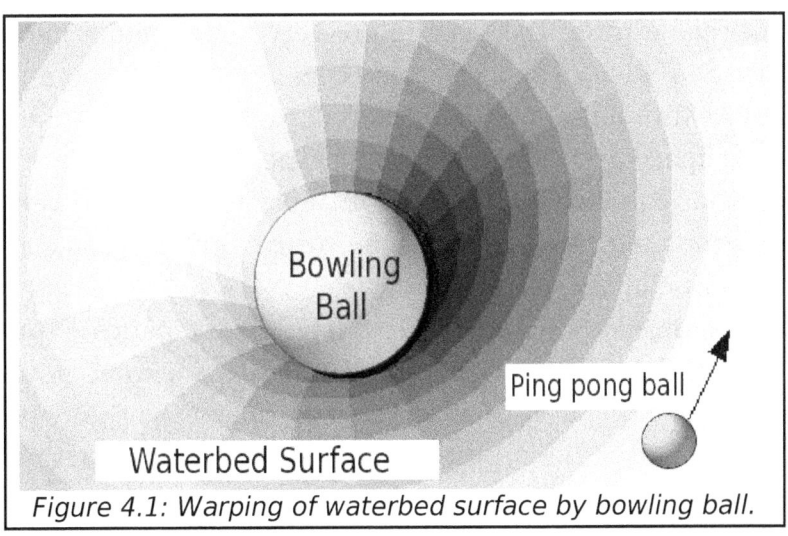

Figure 4.1: Warping of waterbed surface by bowling ball.

According to Newton's #1, the ping-pong ball will continue in a straight line unless another force acts on it. But because the surface is curved, a straight line will be a circle, one of the geodesic curves surrounding the bowling ball. In this example the ping-pong ball will eventually spiral into the center because it is being acted on by both gravity and friction. But if the whole apparatus was put into outer space and the surface was frictionless, the ping-pong ball would circle the bowling ball forever if nothing interferes with it. We think.

Einstein imagined that space is greatly warped near a massive object with the amount of curvature decreasing as distance from the mass increases. The more massive the object, the greater the curvature. All masses have this property but the curvature is only noticeable when the masses are very large (planet-like). Einstein also reasoned that because motion of any object through space cannot be described without time, then space and time must be inextricably entwined. All of the equations for planetary orbital motion were reformulated in a space-time with curved geometry (non-Euclidian).

Besides some truly nasty mathematics, this theory of General Relativity gave rise to some interesting notions. The first is that there is no force of gravity. The motion of the orbiting mass follows Newtons #1 but the curvature of the space makes it *appear* as if a force is present. In his effort to explain the spooky action-at-a-distance force of Newton's gravity, Einstein has introduced some truly spooky properties of mass and space itself. Mass and space are somehow interconnected and mass has a magical property of warping space. Apparently space is not nothing.

I wonder that Einstein wasn't more bothered by his solution than by the original problem, action-at-a-distance. Nevertheless, his General Relativity is widely accepted because it explains anomalies in the orbit of Mercury (perihelion shift) that Newton's gravity theory does not. It also makes predictions that Newton's theory of gravity does not, such as the bending of light around massive objects, which has since been measured.

According to Popper (Chapter 3), we should now abandon Newton's theory in favor of Einstein's. Have we? No. Why not? Because the math is easier in Newton's theory and it does give accurate results in certain limited situations. Only when absolutely necessary do we pull out the big guns and use General Relativity theory. We would probably still be using Ptolemy's geocentric model if the math was easier than Copernicus' heliocentric model. We know it isn't quite right, but if it works then it's useful.

If Einstein's theory of General Relativity is correct, and there is no gravitational force, there only *appears* to be a force from our perspective, then perhaps there are no forces at all. What appears to be a force of nature is an artifact of something else, something more fundamental that we are missing. If so, then Newton's assumption that force is the most fundamental property in nature is in error.

This idea was explored by Joseph-Louis Lagrange (1736 – 1813) and Sir William Rowan Hamilton (1805 – 1865), among others. Using a technique called variational calculus, they assumed that in any kind of motion of physical bodies, a quantity called "action" is minimized. Rather than minimizing the distance an object would travel from point A to point B, for example (a straight line in flat space), they minimized this thing called "action." I have never been able to figure out what this "action" might be. It has units of mass times distance squared divided by time which is equivalent to momentum x distance or energy x time. These units are also the same as Plank's constant (to be discussed in Chapter 6) but Max Plank (1858-1947) came after. Whatever it is, minimizing it leads to Lagrange's equation (the Lagrangian) in one case and Hamilton's equation (the Hamiltonian) in the other formulation.

The entire physics of Newtonian classical mechanics can be reformulated in terms of the Lagrangian. The Hamiltonian, properly defined, is used throughout both classical and quantum mechanics. The principle quantity in both the Lagrangian and the Hamiltonian is energy. Perhaps energy is the fundamental property of Nature.

An aesthetically pleasing feature of the variational method is that it can be formulated in something called "generalized coordinates," which are independent of any special coordinate system. We no longer require a fixed framework called Absolute Space

64

and the space does not need to be flat. The brilliant mathematician, Emmy Noether (1882 – 1935) showed that when transforming the equations of motion from one coordinate system to another, certain quantities remained invariant (unchanging). Using this method she was able to derive all of the physical conservation laws (energy, momentum, angular momentum etc.). Few people understood her paper[1] and it went unnoticed for a long time. This property of invariant quantities being associated with conservation laws is now known as Noether's Theorem.

Have we thrown out Newtonian mechanics in favor of the classical mechanics of Lagrange, Hamilton, and Noether? No. Why not? Because the math is easier using Newtonian mechanics and it gives accurate results in limited situations. Only when needed do we pull out the big guns and use the Lagrangian and the Hamiltonian in classical mechanics.

There is great resistance to change whether or not the math is easier. "The followers of Newton envisaged the Newtonian laws as absolute and universal laws of nature, interpreting them with a dogmatism to which their originator would never have subscribed. This dogmatic reverence of Newtonian particle mechanics prevented the physicists from an unprejudiced appreciation of the analytical principles which came into use during the 18th century, developed by the leading French mathematicians of that period. Even Hamilton's great contributions to mechanics were not recognized by his contemporaries on account of the prevalence of the Newtonian form of mechanics."[2]

Even today, an undergraduate physics student first learns Newtonian mechanics. Lagrange's and Hamilton's methods are generally not taught until the third or fourth year and even then, the Lagrangian and the Hamiltonian are presented without any discussion of historical or philosophical significance. The student barely comprehends what is going on, forget about why.

This chapter on space and time would not be complete if I didn't also mention Einstein's theory of Special Relativity. Einstein was a fan of Mach's Principle, *i.e.* that space is not absolute and the position of a body can only be defined with respect to all of the other bodies in the universe. He was also bothered by the failure of the efforts to detect the aether.

Through imagination (not reason!) Einstein realized that two people will not obtain the same results in measurements of either distance or time if one person is moving relative to the other. Furthermore, the two people will not agree on who is moving, each will think that he is standing still while the other guy is moving.

Einstein noticed that there is no experiment we can perform that will tell us whether or not we are moving as long as we (along with our environment) are moving at a constant velocity (*i.e.* not speeding up or slowing down or turning corners). To convince yourself that this is true, imagine you are in a car traveling along a straight line at a constant speed (on the freeway with cruise control, for example).   If you throw a ball upwards, it goes up and then falls back down just as if you were sitting in your lawn chair.   In fact, if you witnessed someone sitting in a lawn chair next to the freeway it would seem to you as if they were moving past you while you were sitting still. The Earth itself is moving, hurtling through space at an astonishing speed (see box 4.2), yet it seems to you as if you are sitting still calmly reading this book!

Einstein postulated that

1. The speed of light is constant for all observers regardless of who is moving.
2. Nothing can go faster than the speed of light.

These postulates, like Newton's laws, cannot be proved. They are assumed true and we march on from there, another example of a deductive logic system.   Using these postulates people have worked out transformation equations to determine things like length, time and mass measurements from one moving frame of reference to another.

It turns out that the difference between the measurements of length, time and mass are not very different from one reference frame to another unless one happens to be moving close to the speed of light relative to the other. So unless this happens to be the case (as in elementary particle experiments for example), people generally use the simpler classical equations, because the math is easier.

The Earth-shaking thing about Special Relativity is that neither space nor time is absolute. You'd think that variational calculus, General and Special Relativity would have pulled the rug out from under Newtonian mechanics but such is not the case. Every physicist on the planet adheres to Newton's laws. Instead they made up a new rule: in the appropriate limits the new equations have to reduce to the Newtonian equations. This is called the Correspondence Principle. In the case of Special Relativity, in the limit that the velocity is small compared to the speed of light (v << c), everybody measures the same time, distance and mass and Newton's equations are valid.

Italian scientists have just published data from an experiment called OPERA showing that neutrinos can exceed the speed of light.[3] Scientists are rushing to question these results, question the accuracy of the measurement, and looking for alternative explanations for the results. "'If you give up the speed of light, then the construction of special relativity falls down,' says Antonino Zichichi, a theoretical physicist and emeritus professor at the University of Bologna, Italy. Zichichi speculates that the 'superluminal' neutrinos detected by OPERA could be slipping through extra dimensions in space, as predicted by theories such as string theory."[3] When a member of the scientific priesthood makes a remark like this, everyone nods and assumes he is a sage.

Box 4.2 Absolute Reference Frame

If we imagine some absolute (non-moving) reference frame centered at the Earth's core, we can calculate the speed and acceleration we are right now experiencing while sitting in our respective chairs. Using the simplifying assumptions that the Earth is a perfect sphere and we are sitting at the equator, we can approximate our speed by using $v_{us} = d/t$ where d is the circumference of the Earth ($2\pi R_E$) and t is the time taken (1 day = 24 hours). Using $R_E = 6{,}378$ km, $v_{us} = 1{,}670$ km/hr (1,038 mph). The speed of sound at sea level is 1236 km/hr (768 mph) so we are traveling at about Mach 1.4!

Our centripetal acceleration is
$a_{us} = -v_{us}^2/R_E = -437$ km/hr^2 (-0.034 m/s^2) where the negative sign means the acceleration is toward the center of the Earth. Because the Earth is so big, the acceleration is very slight (we don't even feel it).

Now imagine our absolute reference frame at the center of the sun. For this calculation we will assume we are in a circular orbit (it is nearly circular) with the radius being our average distance from the sun, $R_S = 149{,}597{,}870.7$ km (92,955,887.6 mi). The time it takes to complete one orbit is 365.25 days x 24 hrs/day = 8,766 hrs. The velocity of the Earth is
$v_E = 2\pi R_S/t = 107{,}227$ km/hr (66,631 mph)! The centripetal acceleration of the Earth in this orbit is
$a_E = v_E^2/R_S = -76.9$ km/hr^2 (-0.006 m/s^2).

We could then shift our reference frame to the center of the galaxy, but you get the idea. If an absolute frame of reference exists, we can't find it. If we could find it, we are really cooking! But because the acceleration is small (should actually be zero, but we do the approximating thing again) and our entire environment is moving with us, we can't tell. But the Greeks didn't know this so they were scandalized by Aristarchus' claim that the Earth was moving around the sun and not the other way around (Chapter 2).

1. E. Noether, "Invariante Variationsprobleme," Goett. Nachr., pp. 235-257, 1918

2. Cornelius Lanczos, **The Variational Principles of Mechanics**, 4th edition p. 344, Dover Publications, NY, 1949.

3. Geoff Brumfiel, "Particles Break Light-Speed Limit," Published online , Nature, doi:10.1038/news.2011.554, September 22, 2011.

5. Macroscopic Space-Time
(Or The Gospel According to Hubble)

"Houston, We have a problem."
--misquote attributed to James Lovell on the US Apollo 13 flight

There have always been those among us who are fascinated with the stars. Interest was really fueled after Galileo first used the telescope (circa 1609 C.E.) to observe astronomical bodies. How much can we deduce about ourselves, our planet, solar system, galaxy and the entire universe based on observing light from the sky? It's truly amazing how much we know or think we know based on this.

A few hundred years ago, people made up a couple of rules known as the Cosmological Assumptions. Once again, we have to start somewhere, so we make a few rules and see where it takes us. The first assumption is called,

1. The Cosmological Principle: The universe is homogeneous and isotropic.

To say that the universe is homogeneous means that all of the stuff in the universe is evenly distributed throughout. There can't be a whole clump of mass in one corner of the universe and hardly anything

71

everywhere else. To say that the universe is isotropic means that it looks the same in all directions. A cornfield planted in rows is an example of something that is homogeneous but not isotropic. It is homogeneous in that the corn is uniformly spread about and not clumped up in one corner of the field. It is not isotropic because if you look in a direction along the rows it will look different than if you look crosswise to the rows.

The requirement for the Cosmological Principle is a side effect of the Clockwork Rule from Chapter 2. We cannot claim that we occupy a special place in the universe, therefore the universe has to be more-or-less the same everywhere. We got into trouble with the geocentric world view when we made this claim before, so now we have a knee-jerk reaction against it.

We don't really have any evidence in support of the Cosmological Principle and it actually doesn't *seem* to be the case. If you look outside on a clear night there are many more stars in the direction along the galactic disk of the Milky Way than cross-wise to it. Stars are clumped up in galaxies. The answer that the experts give is that it might look that way locally, but if we could get far enough away, then all of the galaxies would be more-or-less evenly spread about and the Cosmological Principle holds.

This would be like saying our cornfield is not isotropic if we only look at one cornfield, but if we look at all of the cornfields, some have rows planted north-south and some are planted east-west so that on a large scale the corn is isotropic. Viewed from an airplane above Iowa maybe.

But the plot thickens. Recently people have mapped out all the known galaxies against what we believe is their relative distances from us and it looks like they are all clumped up. They clump up in a bubble-like or Swiss cheese pattern. They mostly reside on the imaginary surface of gigantic bubbles. But, the experts say, these bubbles or spheres must be evenly spread throughout the universe so at some grand scale the Cosmological Principle holds. You see how it goes. This circular reasoning cannot be trumped by any amount of actual data. But the Earth looks flat from where I am sitting and it doesn't *seem* to be moving. So data can be deceiving, after all. There is something rather suspicious about the Cosmological Principle and it downright gets us into trouble with the Second Law of Thermodynamics as we shall see later in this chapter.

The second Cosmological Assumption is,

2. Universality: The laws of physics are the same everywhere in the universe.

We actually need this one because the only thing we can measure is radiation (usually in the form of light) from the stars. If the laws of physics are not the same over in Alpha Centauri, for example, then we can't infer anything from our measurements of the light that came from there. The mechanisms for the production of the light have to be the same there as they are here.

The Expanding Universe

Why do we think the universe is expanding? It was an accidental discovery made by Vesto Slipher (1875-1969) in 1912 while analyzing data from distant galaxies. In order to understand why we think Slipher's discovery means the universe is expanding, you will need to understand something about how we determine distances in astronomy. Because we can actually measure very little, almost everything we think we know about the cosmos is obtained by inference. Distance determination is a black art which I will attempt to outline.

The distance to even the closest star is vast, about 4.2 light years (LY) away. A light year is the distance light travels in one year (9.46×10^{15} m or 5.88×10^{12} mi). Light from the *nearest* star took over 4 years to get here!

For stars within a radius of about 1500 LY we can determine the distance by triangulation. If we observe an object from two positions separated by a known distance, we can calculate the distance to the object using trigonometry. For these measurements, we use the diameter of the Earth's orbit about the sun. We take the first measurement at position 1 in Figure 5.1, wait six months and take the second measurement at position 2. We also must measure the angle of our telescope, called the parallax angle, designated p in Fig 5.1. The known distance is the diameter of Earth's orbit, which we think we know (see Box 5.1). The accuracy of these measurements is always expressed in terms of the precision with which we can measure the parallax angle, but never the diameter of the orbit. I guess we're pretty confident of that.

Accuracy and precision are not equivalent, though people often use these terms interchangeably. We may be able to measure something with great precision and repeatability because we have a nifty measuring tool, but if our measuring tool is not calibrated, or if there is something faulty in one of our assumptions, our measurement may not be at all accurate.

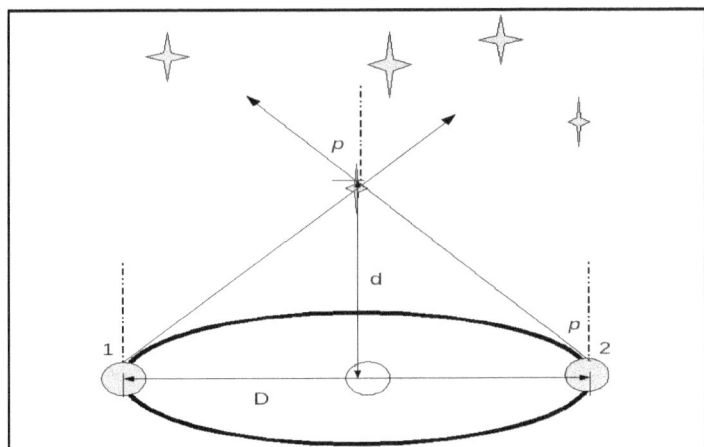

Figure 5.1: Parallax measurement. The sun is not quite at the center because the orbit is not quite circular. Figure is not to scale; D << d.

Box 5.1. Distance from Earth to Sun

It might surprise you to know that we don't actually know the distance from the Earth to the sun. We infer it from the geometry we have invented for our solar system from the observed motion of the planets. There may be other geometries that will work, as evidenced by the geocentric model of Ptolemy (Chapter 2). We can measure the distance from Earth to the nearest planets using radar, but we can't use radar to measure our distance to the sun because the radar just gets absorbed in the sun's atmosphere. "Well," you may say, "we have sent rockets to other planets so our model must be right." To which I say, "Ptolemy's model would have told you which direction to point your rocket in once we knew how far away the planet is. I might have to make minor adjustments in my trajectory along the way, but I'd bet those guys did too." Just because we know where the planet will be and how far away it is does not prove that our overall geometric configuration is correct.

The most precise measurements of stellar parallax to date were taken from a satellite named Hipparcos. From this data we have been able to calculate the distance of stars out to about 1500 LY with a precision of 10% for the nearer stars (about 300 LY). The further the distance, the less precise the measurement. These data only include a small fraction of the stars in our galaxy (thought to be about 100,000 LY in diameter).

To determine distances farther than 1500 LY we need another method. We know that the apparent brightness of a light source depends on distance; *i.e.* the further away it is the dimmer it gets. We can measure the apparent brightness (energy per second per area radiated in our direction). Using that, along with a known distance, we can calculate the actual brightness (total energy per second radiated in all directions). Once we have done this for all of the stars whose distance we measured by parallax we can categorize them and classify them into spectral type (see Box 5.2). We can measure the spectral type of far away stars. We apply the Universality rule: stars over there must be like stars over here. Therefore a star over there of spectral class G say, must have approximately the same actual brightness as a spectral class G star over here. Combining our measured apparent brightness with this assumed actual brightness, we can calculate the distance.

Using this method (called spectral parallax, though it hasn't anything to do with parallax), we have determined distances up to about 33,000 LY with an estimated accuracy of about 25%. To put this into perspective, people doing a controlled experiment in the laboratory try to keep the error around 1-2% or less. If the measurements are particularly difficult, the errors sometimes go up to 5%. Only in the most difficult situations is an error of 10% tolerated. In astronomy and some high energy particle experiments 10% is considered good! A 25% confidence level is terrible, but the astronomers will say that even this ball-park estimate is better than nothing. I agree, so long as we remember that we are dealing with ball-park estimates. In methods used to measure distances beyond 33,000 LY the errors are even greater. People don't even state them anymore. Each of the methods builds on the previous method, so errors or misconceptions anywhere in the process will propagate and multiply.

As an example, suppose I measure the length of your arm and it comes out to 100 cm (1 meter). Suppose my measuring apparatus is only accurate to within 10%. Then I would state my measurement as 100 +- 10 cm. Your arm could be 90 cm or it could be 110 cm or anywhere in between. If my measuring apparatus is only accurate to within 25%, then my stated measurement is 100 +- 25 cm. Your arm is between 75 and 125 cm. At some point it becomes useless information.

Box 5.2 Spectral Class
Another thing we can do with radiation from the stars is to separate its frequency components. You have probably done this with crystals or prisms that break up sunlight into colors like a rainbow. By plotting the intensity of the radiation at each frequency

(called the spectral distribution) we can compare that to known distributions (called Black-Body radiators). From this plot we can infer the temperature of the source. Annie Cannon (1863-1941) developed a spectral classification for stars. From brightest and hottest to dimmest and coolest they are: O B A F G K & M. Our sun is a spectral class G star.

In 1913 Ejnar Hertzsprung (1873-1967) and Norris Russell (1877-1957) independently plotted the total luminosity (or brightness) versus temperature (or spectral type) for all of the stars with known distances (at that time d < 400 LY) and they noticed a pattern; most of the stars were clumped along a curve now called the main sequence. Using this information we developed a method to determine the distances of stars beyond 400 LY. We measure the spectrum of the radiation from a star. From this we determine its temperature and spectral type. If the distant star is on the main sequence we infer its total luminosity from the Hertzsprung-Russell diagram. This, along with the measured apparent luminosity (or brightness) is then used to calculate the distance.

For even greater distances (beyond about 33,000 LY) similar techniques have been devised. We find the brightest stars, Cepheid variables, red giants or supernovae, we find relationships between their actual brightness and distance for those whose distances we measured by some other method, and then we assume that those same type of stars in faraway galaxies have the same actual brightness. With that assumption, along with the measured apparent brightness, we calculate the distances to other galaxies. When the galaxies are too far away to resolve individual stars, we apply the technique to clusters or groups of stars. Eventually we apply it to entire galaxies and clusters of galaxies. You can see that these distance estimates get more and more unreliable.

The City Lights Analogy is often used in astronomy texts to explain this. Suppose you want to measure the distance to all the lights in a city from a rooftop. You are stuck on the roof and must make all of your measurements from there. For the nearest lights you can use triangulation (parallax). By observing from opposite ends of the roof and measuring the length of the roof, you can determine how far away these lights are. Once you know how far away these lights are, you can determine their actual brightness. In the next town over the lights are all too far away to measure their distances by triangulation. However you realize that some of the lights over there are of the same types as those you have already determined over here. Since you know how bright these nearer lights are, you can calculate how far away the lights in the next town are. For yet more distant towns you cannot see individual lights, but you have figured out how far away the next town over is and how much total light it radiates, so you can use this information to estimate how far the distant towns are. At even greater distances only clusters of towns are visible and so on.

I must digress one more time before I tell you about the discovery made by Mr. Slipher. I need to tell you about spectral signatures. Each element in the periodic table has a distinct set of frequencies that it can emit when energy is added. The allowed energies that an atom can absorb or emit are fixed and unique to a specific element. It's like a fingerprint or a signature. Hydrogen has one set of frequencies and helium a different set and so on. This is how we know what elements are present in the atmosphere of the sun and the stars. The visible portion of the spectral signature of hydrogen is shown as a function of wavelength in Figure 5.2 (wavelength and frequency are inversely proportional to each other). Any time we see this pattern we know that hydrogen is present.

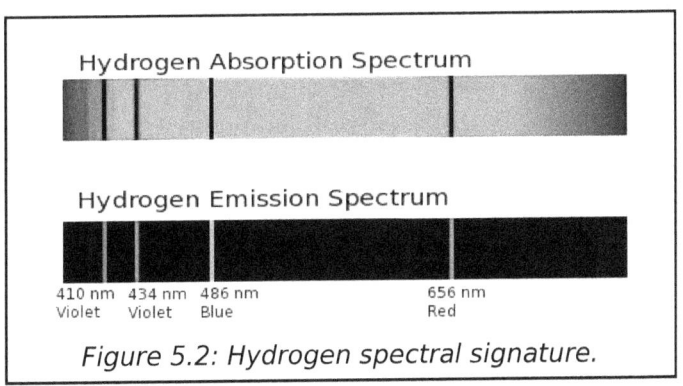

Figure 5.2: Hydrogen spectral signature.

The spectra we see from the stars is like a negative of the emission (bright) spectrum, called an absorption spectrum because the frequencies are actually missing. This is because the stars are radiating energy at all frequencies but the hydrogen gas in the outer atmosphere is absorbing energy, but only the bits allowed by its electronic configuration. The pattern is still present but inverted.

Finally, we get to the discovery of Mr. Slipher (remember him?). In 1912 he was examining the spectra of some distant galaxies to determine their chemical composition when he noticed something strange. Most of the spectra were shifted toward longer wavelengths. The pattern remained intact, it was just moved to the right. Because a shift to the right (to longer wavelengths) is toward the red, this is called a red-shift. A shift to the left (shorter wavelengths), toward the blue is called a blue-shift.

The only explanation we can come up with for this red shift is called the Doppler effect. The Doppler effect applies to waves when the source of the waves is in motion relative to a stationary observer. You might have noticed when a fire truck goes by, the siren has a very high pitch (shorter wavelength when it is coming toward you that lowers (longer wavelength) in pitch as it passes and moves away from you. Because the truck is moving toward you while simultaneously emitting sound waves, the waves get scrunched up. As the truck moves away from you while emitting waves, they are stretched out. The man in Figure 5.3 will hear a higher pitch ("blue-shift") than the woman ("red-shift").

Figure 5.3: The Doppler effect.

If the cosmological red shift is due to the Doppler effect, and if all the objects in the cosmos are moving randomly, we should see blue-shifted and red-shifted spectra about equally. This is true for nearby stars and galaxies but the greater the distance, the greater the red shift until at very great distances all are red-shifted. There is a mathematical relationship between the change in the wavelength and the velocity of the source for Doppler-shifted waves. It seems that no matter which direction we look, if we look out far enough, everything is moving away from us. The farther away the object is, the faster it is moving away from us!

In 1924 Edwin Hubble(1889-1953) and Milton Humason (1891-1972) plotted the velocity versus distance for the known galaxies and discovered a linear relationship (now known as Hubble's Law) as shown in Figure 5.4. This relationship was then used as a tool to determine the distances to galaxies that are so far away their distances cannot be determined by any of the previous methods.

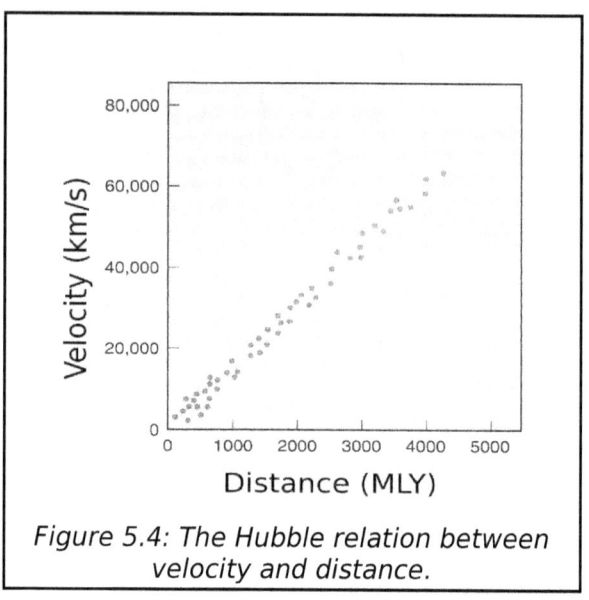

Figure 5.4: The Hubble relation between velocity and distance.

There exist other, more esoteric methods to determine astronomical distances. I have presented here only the primary progression as we try to determine greater distances, summarized in Table 5.1. Each subsequent method builds on the success and accuracy of the previous methods.

Table 5.1. Methods and Range of stellar distances	
Method	**Approximate Range**
Stellar Parallax Angle	< 1500 LY
Spectral Parallax (stars)	1500 – 33,000 LY
Spectral Parallax (star clusters)	33,000 – several million LY
Hubble distance-velocity	> million LY

The Hubble distance-velocity relation is what is known as a Bootstrap Theory, though I have not heard the term in many years. A Bootstrap Theory is when you lift yourself up by grabbing your own bootstraps and pulling. We are using a relationship that we aren't really sure of (red shift is due to velocity of objects), which was established by distance determinations that are very inaccurate, to determine even greater distances. But the situation worsens.

Around 1960 some high intensity radio waves were identified coming from two distant objects named 3C-273 and 3C-48. The spectra from these objects contained numerous lines but they could not be associated with any known chemical element. In 1963, Maarten Schmidt (b. 1929) finally identified the hydrogen spectral lines. They had been shifted so far toward the red that no-one had recognized them until then.

These two objects, now called quasars, have measured red-shifts of 16% and 37% (the percent change in wavelength) which correspond to velocities of

15% and 30% of the speed of light! Since then many more quasars and galaxies with gamma ray bursts have been discovered. Some of these objects have red-shifts indicating they are receding at about 95% of the speed of light ($v = 0.95c$). Friends, this is a problem. This is not allowed by Special Relativity.

There must be some other explanation for the measured red-shift. It cannot mean that these objects are moving because such massive objects simply cannot move that fast. The explanation that has been decided upon, accepted, and handed down from the mountain is this: the objects aren't moving, the space in between is expanding. As an example, take a flat rubber band and paint a wave on it as shown in Figure 5.5a. Now stretch the rubber band as in Figure 5.5b. The rubber band is space (except that space is three dimensional) and the wave is our red-shifted spectral line. Space has some very interesting properties indeed.

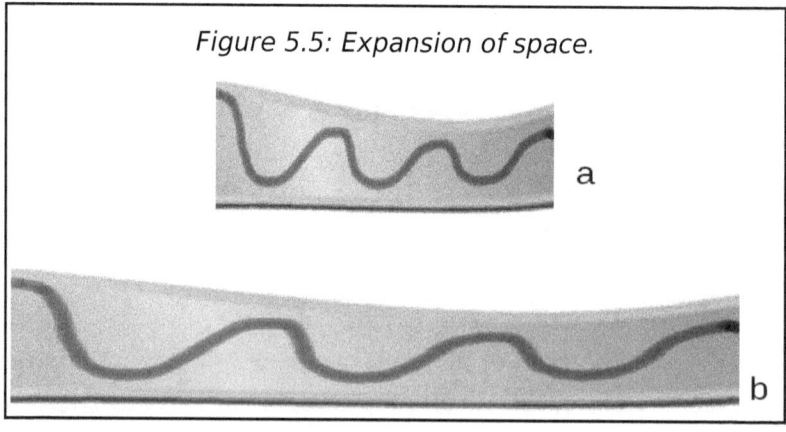

Figure 5.5: Expansion of space.

a

b

The longer our light wave has been traveling through space the more stretched out it becomes. The more stretched out the wave is, the more the red-shift and thus the further away the object that emitted the wave is. Notice that we are not giving up our distance-red-shift relationship. I have a problem with this explanation because, while light (electromagnetic radiation) behaves like a wave, the length of the wave train is ill-defined. It seems to me that the wave would have to be stretched out over a considerable distance for this effect to be noticeable. The wave cannot be affected by whatever the space ahead of it did before it got there and it cannot be affected by whatever space does once it passes by. The only way this expansion of space can affect the wave is if it happens *where the wave also is.*

Even the most coherent laser only has a wave train (coherence length) on the order of a few meters. Your average, garden-variety, white-light wave train is on the order of centimeters. It can even be argued that a photon takes up no space at all. So how can something that takes up no space be affected by whatever strange thing space might be doing? The only way this explanation can have any logic to it is if the light (electromagnetic radiation) does not travel from the quasar to here. Rather it must exist in all of the space between us and the quasar during the whole time space is doing whatever it is doing. In order to make any sense of this we would have to understand what light is. And space. And we do not.

87

An objection that has been raised and addressed is that space does not appear to be doing anything in our local vicinity. We see no change in our relative position within our solar system or our galaxy. The answer offered up by the astronomers is that gravity holds things together so that this expansion of space does not affect gravitationally bound objects: solar systems, galaxies etc.

One little niggling problem is that when we add up all the mass of all the visible stars in any galaxy then we come up short. By a bunch. If gravity is responsible for holding everything together and if mass is the catalyst for gravity, then we need about 95% more of it than we can account for by all the stuff that is visible. But fear not, the astronomers have invented something called "Dark Matter" to solve the problem. Dark Matter is, well, dark. It does not radiate at any wavelength, which is why we haven't seen it. It might be burned-out stars, planets, meteors, dust or some other hypothetical particles. Whatever it is, there has to be a LOT of it. We seem to be missing the components that make up of most of the universe.

The Expanding Universe

If we don't occupy a special place in the universe, then everything in the universe is moving away from everything else. If we run time backwards, everything should then end up at a single point in space. This single point would be the birth of the universe, called the Big Bang. All of the matter and energy present today had to have been created in this Big Bang, since matter and energy cannot be created or destroyed, except, apparently, at the Big Bang itself (another Law that we made up). Matter can be converted into energy and vice-versa. This is the meaning of Einstein's equation $E = mc^2$. Physicists will say that we have no knowledge whatsoever about either the instant of the Big Bang, or the instant before that. Time and space did not exist before the Big Bang, so it is meaningless to talk about it. This is the English translation for: the equations blow up—usually an infinity resulting from a zero in the denominator or some such mathematical abomination.

The priests of the Church like this because, having been trumped by the scientific priesthood for the last few centuries, they finally get the upper hand with the whole creation thing. They all pat each other on the back, say 'good job' and go home.

But there remains a tiny problem. The only way that scientists can know how a system evolves in time is by knowing the initial conditions. For example, if you throw a baseball, I can calculate how far it will fly and where it will land, but only if I know a few things first. I need to know which direction you threw it in and what speed you gave it when it left your hand. These are called "initial conditions."

It turns out that only one set of initial conditions at the instant just after the Big Bang will result in the universe that we see today. We can't have that because that would imply some sort of overall plan. It violates the Clockwork Rule of randomness from Chapter 2. We need to have a plethora of possible initial conditions that lead to the universe today. In order to accomplish this, we invented the Inflationary Universe. This is an extremely short period right after the Big Bang when the universe expanded at an extremely rapid rate—an inflated rate.

Now everything is more or less fine except for one other little problem. The Second Law of Thermodynamics. The Second Law says that the amount of entropy in any closed system must increase with time. This gives a direction to time. We need this because in almost all of the equations in physics, from kinematics to light propagation to elementary particle interactions, it doesn't matter which direction time runs (forward or backward), we get the same answers.

Entropy is a measure of the amount of disorder in a system. The Second Law states that all closed systems move from order to disorder and not the other way around. If a cup falls off the table, hits the floor and breaks, that's okay. It is not okay for the pieces of the cup to weld themselves together and for the cup to jump back up on the table. As long as the cup-table-floor system is closed. If the system is not closed, energy can be added from outside to increase the order (decrease entropy). I can pick up the pieces and glue them back together, for example. But scientists insist that the universe is closed (can't have outside help). If there was somewhere or something outside the universe, inserting energy into our universe, then we would just annex them and our universe is again closed.

So we have a closed universe, a Second Law of Thermodynamics, yet here we are. Except maybe for my desk, things look pretty organized. That we could be put together by random chance from a beginning like the Big Bang is extremely unlikely. It is so unlikely that it can't really happen. Yet here we are. "Well," say the scientists, "It's okay to have more order here in our corner of the universe as long as there is much greater disorder in some other corner of the universe so that the total disorder (entropy) is still increasing." And still we look for intelligent life in the universe.

Box 5.3. Update on 2nd Law of Thermodynamics

One of my reviewers has brought it to my attention that there has been some further controversy surrounding this topic. It seems that the Christians have seized upon this issue to argue for the existence of a creator. They claim that because the probability of our very existence is nearly zero, within this paradigm, then there must have been a divine plan with a mastermind. While I am reluctant to jump in the

---

middle of this one, my only recourse was to delete all references to the 2nd law in this book.

The scientist's response to the Christians was to claim that the density of states at the Big Bang was one. There was only one state. One definition for the entropy S, is

$$S = k \ln(\rho) \quad ,$$

where k is Boltzmann's constant and ρ is the density of states. For $\rho = 1$, $S = k \, ln(1) = 0$ at the instant of the Big Bang. The density of states is steadily increasing because time and space are steadily increasing. In fact, this defines the direction of time. If we allow local fluctuations (more order here so long as there is correspondingly less order somewhere else) then the universe obeys the 2nd law and there is no need for a creator.

Recall that, because of the singularity at the instant of the Big Bang, we can say nothing about it. We can know nothing about that instant, or any instant before it. Time and space simply did not exist. One could just as well argue that there was no density of states because *there was no state*. In that case, the density of states was zero (no state). The natural log of zero is undefined—which is more in keeping with the idea that the Big Bang itself is unknown and undefined. Even if the scientists back off of the instant of the Big Bang and only say that the density of states was small in the early stages of the universe, this will lead us to conclude that the early universe was extraordinarily ordered and most special. This is circular reasoning of the highest order!

Recently, the astronomers have determined that the rate of expansion of the universe is increasing. They believe that the universe is expanding faster now than it was in the past. In order for this to happen, there must be some mechanism pushing everything away from everything else. For this they have invented "Dark Energy." Nobody really knows what that is, but it has to exist everywhere in space and it has to have a negative pressure (it must be repulsive rather than attractive).

Recall from Chapter 2 that the more *ad hoc* insertions we put into a theory in order to fix problems, the more likely the theory is incorrect. I may be the only one who is crazy here, but this emperor has no clothes.

Note: I obtained most of the numbers quoted in this chapter from **In Quest of the Universe,** by Karl F. Kuhn and Theo Koupelis, 3rd edition, Jones and Bartlett, 2001. There is a wide range of reported values for the range and accuracy of astronomical distance measurements. Those quoted here were obtained online from Wikipedia.

6. Microscopic Space-Time
(Or The Gospel Concocted in Copenhagen)

"No point is more central than this, that empty space is not empty. It is the seat of the most violent physics."
--Misner, Thorne & Wheeler, **Gravitation**, W.H. Freeman and Co., San Francisco, p.1202, 1973

One of the central ideas of quantum mechanics is that nothing in our reality seems to be continuous. Suppose we draw a line and label one end zero and the other end one. Our line looks continuous, but if we start dividing up our line into smaller and smaller pieces, we will eventually get to a place where it can be divided up no more. Turns out that everything in the universe, length, momentum, energy etc. can only be divided up so far. The smallest chunk we can divide things up into is called a quanta and, since Max Plank (1858-1947) first postulated this idea, there will be a constant associated with each type of chunk called Plank's constant, h ($h =$ 6.626 x 10^{-34} kg m^2/s). This was quite a surprise because everything *seems* to increase or decrease smoothly and continuously. But because Plank's constant is so small, the chunks are small and they are beneath our perception. At the microscopic level, these chunks are important and they comprise the theory of quantum mechanics. (A "quantum leap" by the way, is *very* tiny.)

This whole quantum thing came about because of what is known as the **ultra-violet catastrophe**. The Rayleigh-Jeans law that describes the intensity versus frequency of black body radiation breaks down at high frequencies, *i.e.* in the ultra-violet. A black body is a perfect absorber; it absorbs all radiation incident upon it. When a black body is in thermal equilibrium with its environment, it re-radiates all of the radiation that it absorbs. So it just sits around absorbing and re-emitting radiation. The experimental curve of intensity versus frequency can be used to determine the temperature of the black body and its immediate surroundings. This is the way that the temperature of stars is measured, as discussed in Chapter 5. Max Plank came along and assumed that energy can only exist in chunks, the smallest chunk of electromagnetic energy, he decided, was $E = hf$, where h is Plank's constant, and f is the frequency of the radiation. He plugged this into the distribution of energy from thermodynamics and got the right answer (after determining h from the experimental black body curves). Even he didn't understand why.

After Max Plank did this *ad hoc* quantization people started quantizing everything, but they did it (and still do) in this same *ad hoc* way, without really understanding what is going on. I traced the calculations that John William Strutt, third Baron

Rayleigh (1842-1919) and James Jeans (1877-1946) did to arrive at the Rayleigh-Jeans law of black body radiation and I found an approximation that explains why they got the wrong result. When calculating the energy in a black body cavity they assumed that the cavity is large compared with the wavelength of the radiation and therefore the energies are smooth and continuous. Ordinarily this would seem to be a good approximation if your cavity is, say a foot or two in diameter (or more) because the typical wavelengths are on the order of microns and nanometers. Compared to a foot, what's a couple of nanometers?

It so happens that the allowed solutions for the radiation inside the cavity were already quantized. When Rayleigh and Jeans made this approximation, they did away with the quantization **that was already there**. You are probably familiar with waves on a stringed musical instrument. When plucked, the string vibrates, but only at certain frequencies that are characteristic of the material the string is made of, the tension in the string and, most importantly, the length of the string. The lowest frequency (and thus energy) at which the string can vibrate is when the wavelength, λ (inversely proportional to frequency) is exactly equal to twice the string length, L. The next highest frequency is when $\lambda = L$ and so on. For resonance (music) to occur, the only allowed wavelengths are $\lambda = 2L/n$, where n = 1, 2, 3, 4... The first three resonances are shown in Figure 6.1.

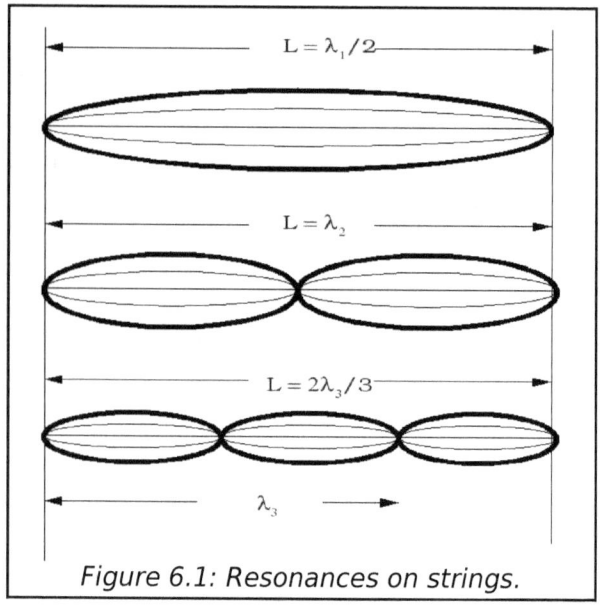

Figure 6.1: Resonances on strings.

Now if we cut the string in half, we no longer have the lowest mode that we had before, but we still have all the higher harmonics. Our lowest harmonic frequency (energy) is now twice what it was before we cut the string in half. If we continue to cut the string in half we continue to double our lowest harmonic frequency (while cutting the wavelength in half). But how many times can we cut the string in half and still hear music? Something has to vibrate to make the sound. There is some smallest length of string that will support a resonant vibration. In the case of our black body cavity, Raleigh & Jeans were dealing with harmonics inside the cavity with a very large value for n, the quantum number. In analogy with the string, they ignored the very high frequency modes where the wavelength is *much* smaller than the string length. But it is exactly at the high frequencies (short wavelengths) where the Rayleigh-Jeans law breaks down! They did not ask themselves what happens when the cavity gets small. As with the string, there must be some smallest size of cavity that can support a resonant wave.

We think that electrons jiggle around and produce electromagnetic radiation. The smallest cavity that can support a resonant wave must therefore have a jiggling electron in it. The size of the cavity simply cannot go to zero (which is really what the assumption means; everything can smoothly go to zero: the cavity size, the energy, everything). I redid this calculation assuming that the smallest cavity size supports one oscillating electron and I was able to derive Plank's constant. The details of the calculation are in Box 6.1. I think all of nature is naturally quantized in this way. The quantization is a result of imposing boundaries on energy.

Box 6.1 Derivation of Plank's Constant

In David Bohm's (1917-1992) excellent textbook on Quantum Theory,[1] he devotes the entire first chapter to the derivation of the Rayleigh-Jeans law and Plank's hypothesis. For details, please refer to his original text. Bohm first derives the vector potential

---

everywhere inside a cubic black body cavity with non-conducting walls. The equilibrium distribution of energy density in a hollow cavity is independent of the shape of the container so he chose the cube because it gives the simplest solutions. The wave equation can be derived from Maxwell's equations in terms of the vector potential, $\boldsymbol{a}$. The standard solution to the wave equation with periodic boundary conditions is

$$a=\sum_{l,m,n} a_{l,m,n}\cos(k_{l,m,n}\cdot r)+b_{l,m,n}\sin(k_{l,m,n}\cdot r)$$
(6.1)

where l, m, and n are integers from $-\infty$ to ∞ and the bold characters indicate the quantity is a vector. The simplest geometry is a cube that is arbitrarily divided up into smaller cubes, each with sides L. The boundary condition is that the field must be the same at corresponding points of every cube. This leads to the solution given in Eq. (6.1) with the wave vector

$k_{l,m,n}^2=(\frac{2\pi}{L})^2(l^2+m^2+n^2)$. The magnitude of the wave

vector $\boldsymbol{k}$ is $2\pi/\lambda$ where λ is the wavelength. The resonant waves are those that exactly fit into each

cube, i.e. $\lambda=\dfrac{L}{(l^2+m^2+n^2)}$.

The electric field is given by

$$\varepsilon=\frac{-1}{c}\frac{\partial a}{\partial t}$$

$$=\frac{-1}{c}\sum_{l,m,n}\dot{a}_{l,m,n}\cos(k_{l,m,n}\cdot r)+\dot{b}_{l,m,n}\sin(k_{l,m,n}\cdot r)$$
(6.2)

where $\dot{a}$ and $\dot{b}$ are time derivatives. The magnetic field is

$$H = \nabla \times a$$
$$= \sum_{l,m,n} -k_{l,m,n} \times a_{l,m,n} \sin(k_{l,m,n} \cdot r) + k_{l,m,n} \times b_{l,m,n} \cos(k_{l,m,n} \cdot r). \qquad (6.3)$$

The energy density of an electromagnetic field is

$$u = u_\varepsilon + u_H = \frac{\epsilon_o \varepsilon^2}{2} + \frac{H^2}{2\mu_o} , \qquad (6.4)$$

where ϵ_o and μ_o are the permittivity and permeability of free space, respectively. The total energy in a cavity of side L is the integral over the cavity volume of the energy density,

$$E = \int_V u \, dV$$
$$= \frac{V}{2c^2} \sum_{l,m,n} \epsilon_o [(\dot{a}_{l,m,n})^2 + (\dot{b}_{l,m,n})^2] + \frac{c^2 k_{l,m,n}^2}{\mu_o} [(a_{l,m,n})^2 + (b_{l,m,n})^2] \qquad (6.5)$$

which is Eq. (23) in Bohm's **Quantum Theory**.[1] From here Bohm proceeds to calculate the number of oscillators in the cavity. He then combines that result with the average energy per oscillator, given by the equipartition theorem, to arrive at the Rayleigh-Jeans law. When calculating the number of oscillators, he says: "Now, for any reasonable value of **k**, the number of waves fitting into a box is usually very large. For example, at moderate temperatures, most of the radiation is in the infrared, with wavelengths ~10^{-4} cm. Hence, when **k** changes in such a way that one more wavelength fits into the box, only a very small fractional shift of **k** results. It is possible, therefore, to choose the interval d**k** so small that no important physical quantity (such as the mean energy) changes appreciably within it, yet so large that very many radiation oscillators are included. This means that the number of oscillators can be treated as virtually continuous, so that we can represent it in terms of a density function."[1] And this is where I part company.

The problem with the Rayleigh-Jeans law is in the limit of small wavelengths (high frequency). We cannot assume that the cavity size, which is directly proportional to the wavelength, can go to zero smoothly. There is some smallest size we can have that will still support one resonant mode. If we assume, as we do, that electromagnetic radiation results from oscillating electrons, then the smallest cavity must support a resonant vibration for at least one electron.

It is well known that the energy density of an electromagnetic wave is equally shared by the electric and magnetic fields, *i.e.* $u_\varepsilon = u_H$, therefore $u = 2u_\varepsilon = 2u_H$ and

$$E_{l,m,n} = \frac{V}{2\mu_o} \sum_{l,m,n} k_{l,m,n}^2 [(a_{l,m,n})^2 + (b_{l,m,n})^2] \quad . \tag{6.6}$$

We want to calculate the energy for the lowest possible mode. Each mode can have 3 possible polarizations (x, y, or z), corresponding to the possibilities for the vector, $\mathbf{k}$. (The lowest energy state is degenerate since $E_{1,0,0} = E_{0,1,0} = E_{0,0,1}$.) So the lowest energy is $E = E_{1,0,0} + E_{0,1,0} + E_{0,0,1} = 3E_{1,0,0}$. We have an additional restriction on the boundary; the field has to be zero at the walls, therefore the $a_{l,m,n}$ are all zero (the coefficients of the cosine in Eq. (6.1)). The lowest possible energy is then,

$$E_{min} = \frac{3V}{2\mu_o} k_{1,0,0}^2 b_{1,0,0}^2 \quad . \tag{6.7}$$

It can be shown that the vector potential along the axis of a filament current loop of radius $\mathbf{R}$ is,[2]

$$b = \frac{\mu_o I}{2} \frac{R}{\sqrt{(z^2 + R^2)}} \hat{\phi} \quad , \tag{6.8}$$

where $\mathbf{b}$ is in a direction along the filament $\hat{\phi}$, I is the current in the filament, $\rho = \sqrt{(z^2 + R^2)}$ is the distance from the filament to the z-axis. The average value of $\mathbf{b}$ is

103

$$<b> = \frac{1}{2R} \int_{-R}^{R} b\, dz = \frac{\mu_0 I}{2} \int_0^R \frac{dz}{\sqrt{(z^2 + R^2)}}$$

$$= \frac{\mu_0 e \omega}{2} \ln(1 + \sqrt{(2)}) = \mu_0 \pi e f \ln(1 + \sqrt{(2)})$$

$$(6.9)$$

for a single electron, where e is the charge of an electron and f is its oscillation frequency. Combining this result with the lowest energy in Eq. (6.7) gives

$$E_{min} = \frac{6\pi^4 V}{\lambda_{1,0,0}^3} \mu_0 e^2 c f \ln^2(1 + \sqrt{(2)}) \qquad (6.10)$$

where I have used $k = 2\pi/\lambda$ and $f = c/\lambda$. The lowest allowable mode in the cavity is $\lambda = 2L$, along with $V = L^3$ results in,

$$E_{min} = \frac{3}{4} \pi^4 \mu_0 e^2 c f \ln^2(1 + \sqrt{(2)}) \quad . \qquad (6.11)$$

With c $= 3 \times 10^8$ m/s, e $= 1.6 \times 10^{-19}$ C, and $\mu = 4\pi \times 10^{-7}$ kg m/C^2 ,

$$E_{min} = 5.5 \times 10^{-34} (J \cdot s) f \approx hf \quad . \qquad (6.12)$$

The currently accepted value for Plank's constant is $h = 6.626 \times 10^{-34} J \cdot s$. Clearly, this calculation is an approximation. I think the biggest discrepancy has to do with the coordinate systems. The vector potential from the wave equation was calculated in a rectangular coordinate system (x, y, z) and I used a cylindrical coordinate system (ρ, ϕ, z) to estimate the vector potential for a current loop. I think that cavity shape does matter at this scale. Because rotating and spinning seem to be natural motions, I suspect one has to go recalculate everything in spherical coordinates. I will let someone else do this because I don't like Bessel functions.

In addition, the value of these physical constants are not usually directly measurable, but are inferred and usually depend on the values of other constants. These values change over time depending on how they are being measured.[3]

Even though this calculation is an approximation, the fact that it comes so close to predicting the value of h is striking. The central point of my argument is this: nature is automatically quantized any time the energy is confined (*i.e.* boundary conditions are imposed). In fact, the definition of mass is confined energy according to Einstein's equation $E = mc^2$.

Atomic Structure

You probably think we know what an atom looks like. We do not. Every thing at the atomic scale and smaller is measured by inference. We see a voltage spike from our detector and, depending on what type of detector it is, we infer the presence of an electron or a photon or an xyz-on. I call them xyz-ons because, every time we have a problem, we invent a new xyz-on to solve it. Then we go looking for an xyz-on and usually we find one. I can't tell if this has anything to do with reality or not, but this is the process.

We have a model for the atom (the Bohr model) but we think that it isn't right. We still teach it in our schools because we don't have another one and the Bohr model is useful within certain limits. Because we still teach a model that we know is incorrect, most of us can't get the idea of an atom as a miniature solar system out of our heads.

We think an atom has a hard core at its center because we can bounce other particles (neutrons and such) off of it like billiard balls. We call this hard core the nucleus. It has properties we call mass and charge, though we don't really know what those are. Hanging around the nucleus we have this cloud of energy stuff that we call electrons. This cloud of electrons is oppositely charged from the nucleus and can only have certain configurations. We don't know why, but we made up a bunch of rules that work (quantum selection rules) to predict atomic spectra and such like.

But when we really get right down in there and try to figure out what an xyz-on is, the dang thing disappears on us. In fact, it doesn't seem to be a thing at all. It seems to be a no-thing that is no-where. Until we look for it. In the process of looking for it (detection), we destroy whatever state it *was* in and we can no longer say anything about the state it *is* in or *will be* in.

This was a big shock when it was discovered in the early 20th century.

Since then quantum theory has passed every test human ingenuity can devise, to an amazingly accurate degree. However, physicists are at a loss to explain the meaning of the theory. Science used to start with the physical explanation, based on data, and develop the theory from the understanding of the phenomenon. With quantum mechanics, the situation is quite reversed: we have the theory and no physical explanation. In fact, because of the fantastic successes of quantum mechanics, science has ventured far into a land of mathematical abstractions, with little hope of explanations.

"Quantum theory resembles an elaborate tower whose middle stories are complete and occupied. Most of the workmen are crowded together on top, making plans and pouring forms for the next stories. Meanwhile the building's foundation consists of the same temporary scaffolding that was rigged up to get the project started. Although he must pass through them to get to the rest of the city, the average physicist shuns these lower floors with a kind of superstitious dread.

"... Physicists' reality crisis consists of the fact that nobody can agree on what's holding the building up. Different people looking at the same theory come up with profoundly different models of reality, all of them outlandish compared to the ordinary experience which constitutes both daily life and the quantum facts."[4]

Quantum Mechanics and the Wave-Particle Duality

There are at least three unexplained enigmas associated with quantum theory. The first of these is called the **wave-particle duality**. Before quantum mechanics came on the scene, physicists had the world neatly divided up. Light and sound were waves and electrons and atoms and such were particles. Each has a whole set of associated equations. It isn't appropriate to apply the wave equations to a particle and vice-versa. For starters, a wave doesn't have the properties of mass, charge and so forth. A wave has a well-defined velocity and wavelength (distance from crest to crest) but it does not have a well-defined position; it exists over a whole area (think of waves on the ocean). A particle has a well-defined position; we say it is localized in space. In the case of a macroscopic particle, like a baseball, recall from Chapter 5 that we need to specify the initial conditions, velocity and position, in order to determine where and when it will land.

In the early part of the 20th century, it was noticed that light was acting like a particle. It was knocking electrons out of a metal surface just like little billiard balls bouncing into each other (now known as the photo-electric effect). So the physicists invented a light-particle and called it a photon.

In 1924, Louis de Broglie (1892-1987) got the bright idea that, since light exhibits particle-like behavior, maybe particles (specifically electrons) exhibit wave-like behavior. So people set up wave-like experiments (interference) with electrons and, sure enough, they acted like waves. Since then all sorts of particles have been shown to have wave-like properties, even very large molecules. Probably baseballs do too. People still don't know what to make of this because the two concepts, waves and particles, seem mutually exclusive.

The wave-particle duality is a manifestation of the Uncertainty Principle of quantum mechanics. Mathematically, the uncertainty principle is

$$\Delta x \, \Delta p \geq \hbar \tag{6.13}$$

where Δx is the minimum uncertainty in position x, Δp is the minimum uncertainty in momentum p (momentum = mass x velocity), and $\hbar$ is Plank's constant divided by 2π. What it means is that if we force Δx to be very small (we know the position of a particle exactly) then Δp must get very large (we don't know its velocity at all). Conversely, if we know the velocity exactly, then Δp is very small but then Δx must get large and we don't have any idea where the particle is located. The latter is characteristic of a wave, whose momentum is well-defined but is everywhere in space.

This pretty much messes up our whole program because we can no longer specify the initial conditions and therefore we cannot make any predictions about where our particle will end up later. This is why we are reduced to knowing only probabilities when it comes to microscopic particles. We can only state the probability of finding the particle here or there, or the probability of the particle having a certain velocity. Our equations of motion are replaced by a wave function; a sum of possible states with varying probabilities. That's the nature of the game in the microscopic world, eliciting the famous statement, "God does not play dice!" from Albert Einstein.

Now, a wave has a well-defined momentum but it is spread out over space. This corresponds to Δp small and Δx large. A particle is confined in space corresponding to Δx small, therefore Δp must be large. Whatever a particle is, it is moving around and changing direction inside its confined volume. The smaller the volume in which we confine it, the crazier its motion gets. Thus the wave-particle duality is associated with the uncertainty principle. Whether or not these whatever-they-are are waves or particles depends on our choice of measurement.

Quantum Mechanics and The Measurement Problem

The second unexplained enigma associated with quantum theory is known as the **measurement**

problem. In the world of classical physics, a baseball (or any object) has a well-defined state. The state will have an associated energy and momentum. An elementary particle, say a proton, does not. It exists in what is called a superposition of states; that is, it exists in many states at once. We don't know what state it is actually in until we measure it. Then it "magically" settles down into a single state, the one we measure (except that our measurement has destroyed the state). It is not correct to assume that the proton was in that state before we measured it and we just didn't have enough information to know which state it was actually in. It was literally in all of those states, or none of them, before we measured it. Thus, the theory that describes the motion of elementary particles is a statistical one, comprising all possibilities.

The mathematical entity that contains this statistical information for all possible states is called the wave function. The wave function is a sum of each possible state multiplied by a quantity that is associated with the probability of finding the particle in that state. When a measurement is made, the wave function "collapses" into a single state. There is no mathematical operation or transformation for which this can naturally occur. This puts the measurement itself into a very special position. Questions can then be asked about the measuring device versus a real human being (*i.e.* does the wave function collapse when the device makes the measurement, or when the scientist reads the measurement from the device?). This is very different from classical Newtonian certainty. There have been many interpretations of what this might mean about the nature of elementary particles and, ultimately, reality.

The various interpretations are as follows.

Interpretation #1: **The Copenhagen interpretation**, there is no deep reality (so-called because of a meeting held in Copenhagen in 1941 to discuss this topic).

Probabilities are all we can know and there is nothing more to be done about it. Elementary particles exist, but have no intrinsic properties before we measure them and cannot be considered "things" in any real sense. There is no hidden meaning. This interpretation was proposed by Neils Bohr (1885-1962) and Werner Heisenberg (1901-1976). This is the most widely held belief among physicists. It doesn't really explain anything, but states that explanations are unnecessary; we have all the tools we need to do the calculations and carry out experiments.

Interpretation #2: **Observation creates reality**.

Elementary particles (and thus reality) do not exist until their properties are measured. Reality does not exist until we perceive it. In perceiving it, we are creating it. "Wheeler takes observer-created reality a step beyond ...with what he calls a 'delayed-choice experiment'. In such an experiment, the observer creates not only present attributes of quantum entities, but also attributes that such entities possessed far back in the past, which by conventional thinking existed long before the experiment was conceived, let alone carried out..."[4] This implies that our observation of the state of the particle not only creates it, but also creates its entire history.

Interpretation #3: **Consciousness creates reality.**

This differs from the second interpretation in that, in Interpretation #2, anything can be the observer, any kind of animal (person) or machine (computer). In Interpretation #3, only consciousness has the capability to make something real. This logic was first proposed by Jon Von Neumann (1903-1957), where he insisted that measuring devices cannot have a special place in the universe. The only alternative is that the measurement *act* must be special. Only the entity performing the *act* of measurement therefore has the power to collapse the wave function. In this interpretation, the world is not objectively real but depends on the mind of the observer.

Interpretation #4: **Hidden variables; wholeness and the implicate order**.

David Bohm and Albert Einstein were very disturbed by the statistical nature of quantum mechanics. Bohm insisted that there must be some hidden variables that are not apparent in the theory. If we knew the nature of the hidden variables, then the theory would be complete and we could calculate with certainty the state of an elementary particle. From this he developed the concept of the implicate and explicate universe. The implicate universe is where the hidden variables are, the explicate is what we experience. In this theory, the entire universe is connected through the implicate order, therefore all aspects of reality (the whole universe?) must be taken into account for any given measurement in order to achieve a unique solution (a very Machian idea).

Interpretation #5: **The many-worlds interpretation**.

This interpretation was proposed by Hugh Everett (1930-1982) in 1957 for part of his PhD dissertation. Everett decided that there can be nothing unique or special about measuring devices. He therefore postulated that everything that can happen, does. The wave function does not collapse, but each possibility is realized, each in its own separate reality or world. He calls these parallel universes, presumably because two parallel lines will never meet (in flat space-time) and these universes also never meet since we are only aware of one measurement.

"Everett's many-worlds interpretation of quantum theory, despite its extravagant assumption of numerous unobservable parallel worlds, is a favorite model of many theoretical physicists because of all quantum realities it alone seems to solve the measurement problem with no arbitrary canonization of the process of measurement.

"Einstein objected to suggestions of observer-created reality in quantum theory by saying that he could not imagine that a mouse could change the universe drastically simply by looking at it. Everett answers Einstein's objection by saying that the actual situation is quite the other way around. 'It is not so much the system,' Everett says, 'which is affected by an observation, as the observer who becomes correlated to the system.'"[4]

Interpretation #6: **Quantum logic**.

The quantum world obeys a logic that is non-native to humans. If we could find the point of view of the quantum stuff, everything would make sense.

Though it has been debated for almost a hundred years now, a resolution to the measurement problem has still not been agreed upon. Most physicists accept the Copenhagen interpretation and don't worry about it anymore. But no explanation is not an explanation. If the point of physics is to explain reality then this is a very serious cop-out. Observation creates reality and Everett's many worlds interpretations have gotten a lot of press, but these are the least favorite explanations for most physicists.

Quantum Mechanics and Non-Locality

The third enigma of quantum mechanics is **non-locality**. In 1935, Einstein, Boris Podolsky (1896-1966) and Nathan Rosen (1909-1995) made up a thought experiment now known as the EPR paradox. Their purpose was to expose how ridiculous the whole idea of quantum mechanics is. Their thought experiment utilizes a property of the electron called spin. The electron spin can have two possible values, called up and down for simplicity. We can prepare two electrons in the laboratory such that their spin states are correlated: if electron #1 has spin up, then electron #2 has spin down and vice-versa (this is a consequence of conservation of angular momentum). Each electron is *not* in a definite spin state, but rather a superposition of both the up and down states. The spin of electron #1 is *neither up nor down until we measure it*. The same is true of electron #2. Once we measure the spin state of one of them, the other one will have the opposite state. These are called entangled states.

Now, suppose I keep electron #1 with me in the laboratory and send electron #2 to the moon along with my collaborator. (Remember, this is a thought experiment.) The spin states of the electrons are still undetermined and still entangled. At some pre-determined time, I and my collaborator simultaneously measure the spin states of our respective electrons. If I measure spin up, my collaborator measures spin down.

If I measure spin down, my collaborator measures spin up. Every time.

The question is this: if neither electron was in a fixed spin state before the measurement, then how did electron #2 on the moon know what my measurement of the spin of electron #1 was and vice-versa? The information was somehow instantaneously passed (called super-luminal communication) faster than the speed of light—and we've already discussed that!

Since Einstein, Podolsky & Rosen published the now-famous EPR paradox, many experiments have been performed verifying the validity of their joke. The net result seems to be that elementary particles are somehow connected to their environment and to each other. Changes in the environment are transmitted instantaneously to all. No one knows what to make of this.

An Embarrassment of Theories

By now I hope you have noticed that we apply a different theory for just about every situation we encounter. This is why we have so many specialists; someone who is working in lasers and semi-conductors has no idea what is going on in astrophysics and vice-versa. Each scientist is off in her little corner and the science in that corner is so challenging that she doesn't have time to think of anything else. The only folks who are keeping track of the big picture are maybe the philosophers and, as I already mentioned, no-one listens to them anymore.

This plethora of theories has bothered some theorists. Not only is it an embarrassment, it is very ugly. A beautiful theory would be One Theory that predicts everything. We would also like it to reduce to all the sub-theories that we know and love. Some really smart people have been working on this for a fairly long time (~75 years) and have not succeeded.

The first thing the theorists did was to gather up all the forces: electromagnetic, gravitational, strong nuclear, and weak nuclear. Everything we know can be described mathematically in terms of these. You might be wondering what happened to Einstein's idea that gravity is not a force, but only looks like one from a certain perspective. Well, one can make the same argument for the electromagnetic force. The electromagnetic force arises from the charge of a particle (like mass with gravity). One could say that the charge warps the space near the particle but with a shorter range than that of mass. It's not clear how one would handle the other two forces. But the math is *really* hard. Either that or they just don't want to give up Newton's idea of forces being paramount, I'm not sure which.

The second thing the theorists did was to make up a field to go with each force. The fields and the forces are connected mathematically. Nobody quite knows what a field is and one of my former professors swears that they don't exist. Nowadays many physicists believe that *only* fields exist, particles being an excitation or a condensation of the field at a specific location. In general, fields are invisible and undetectable. We infer their presence by how a particle behaves when it is in one (the force it experiences). A field is a quantity that depends on position in space and time. In one viewpoint, the field is created by the particle and this can explain "spooky action-at-a-distance." The action is perpetrated through the induced field.

The third thing the theorists did was to quantize the fields. They did this because we have figured out that everything in nature is quantized: charge, mass, energy, everything. So we suppose the fields must be too. In the case of the electromagnetic field this was a very successful approach, now known by the intimidating name of Quantum Electro-Dynamics (or QED). They hit a snag early on though. Every time they tried to calculate something they got infinity for an answer. In calculating the probability for an electron to go from point A to point B, for example, one has to take into account every possible path and every possible interaction that the electron can encounter on its way. What happens is that when you take the calculations for interactions down to zero distance between interactions, the equations blow up. This is a consequence of the uncertainty principle.

Richard Feynman (1918-1988), among others, made up some rules to deal with this problem. They called it *renormalization*, which amounts to subtracting off infinity or figuring out ways to ignore it. "Having to resort to such hocus-pocus has prevented us from proving that the theory of quantum electrodynamics is mathematically self-consistent one way or the other by now; I suspect that renormalization is not mathematically legitimate."[5] Why we don't hear the mathematicians howling is a mystery to me. Despite this, renormalization has turned into a test of a good theory! If it *can* be renormalized, it is considered a good theory because it turns out that many theories can't.

Amazingly enough, this seems to work remarkably well in the case of QED, though it has some rather strange features. Particles can move either forward or backward in time; it makes no difference mathematically. An electron going backward in time is a positron and vice-versa. The mechanism they invented for the force between charged particles is called "virtual particle exchange." It's not clear how this works, but the virtual particle for the electromagnetic force is the photon (light). When a physicist says "virtual," she means it doesn't exist. It is not real. These virtual particles cannot be real because if they were they would violate the conservation laws. Can't have that; it's against the law. As long as these virtual particles don't exist for very long, everything is copacetic. We can violate the conservation laws if we do it very fast.

Particles and particle-antiparticle pairs can be spontaneously created out of nothing (the vacuum)! In fact, the vacuum is called the zero-point energy. The infinity that we got rid of is considered to be the energy of the vacuum. That is, empty space has infinite energy! Seems like nothing (space or the vacuum) is something and something (particles and such) are nothing.

Because of the great success of QED, the same technique was tried for the other three fields: gravity, strong and weak nuclear. If we could successfully do this we would have our ONE theory, Field Theory. Stephen Weinberg (b. 1933), Abdus Salam (1926-1996) and Sheldon Glashow (b. 1932) successfully combined QED with the weak nuclear interactions and won a Nobel prize for it in 1979. The virtual exchange particle is called the W-particle.

For the nuclear force, the theorists invented quarks, each nucleon being a combination of three quarks, with a virtual particle called the gluon holding them together somehow. This one is called Quantum Chromo-Dynamics (QCD) but it wasn't so successful because it turns out that it is not so easily renormalizable. One can't ignore a whole bunch of mathematical terms as in QED, and these non-negligible terms are extremely difficult to calculate. For the gravitational field, the theorists invented the virtual graviton as the exchange particle, but this theory was an utter failure.

We ended up with a field theory only useful for electromagnetic and weak nuclear interactions, called the electro-weak theory. For the nuclear force, theorists moved on to what is called the Scattering Matrix, or S-matrix.

When protons and neutrons are involved in high energy collisions, all sorts of exotic particles emerge: mesons and baryons of various energies. All of these are made up of quarks and as such are members of a family of particles called hadrons. Each of the hadrons are now thought of as being excited (higher energy) states of the basic hadron particles, protons and neutrons.

The S-matrix is a collection of probabilities for all possible interactions involving hadrons. In S-matrix theory, only the initial and final states of the interactions are specified; the mechanism (virtual particle exchange for example) is not. The S-matrix theory does not have the renormalization problem that field theory has because the momentum is fully specified but position is not. In the S-matrix picture, particles are seen as events rather than things. They are called energy resonances. A resonance is a large blip in energy.

The S-matrix theory seemed promising but problems were encountered when the theorists insisted on embedding certain principles into it: results must be independent of position in space, time and motion, and cause and effect must be preserved (defining a direction for time). They were unable to make this work even for the strong nuclear force, let alone gravity and the electro-weak.

The S-matrix is an abstract mathematical object known as a group. There are many such objects. So when the S-matrix didn't pan out, the theorists tried some of these other groups. In order to make it all work, they had to keep switching to groups of higher and higher dimensions. Eventually they landed on a group called a string. In this picture the particle resonances are oscillations of the strings. Just like the string on a guitar, only certain harmonic modes are allowed, corresponding to the observed particles.

It turns out there are many groups of strings with various dimensions (10, 11, 26) that might fit the bill. Some are related to each other through various mathematical transformations. The search for the Theory of Everything is still on. Some of the higher-dimensional groups, called "branes" (a string is a higher dimensional object than a point, a membrane is a higher dimensional object than a string and so on) are thought to be candidates.

Superstring theory is a possible unified theory of all fundamental forces, but superstring theory requires a 10 dimensional space-time. The problem is how to reduce these 10 dimensions to the 4 (3 space, 1 time) dimensions of the physical world. One proposal is to roll up the extra dimensions into some very tiny but nonetheless interesting space of their own that is not perceivable by us. Another suggestion is to make the extra dimensions really big, but constrain all the matter and gravity to propagate in a three dimensional subspace called the three-brane. This page of your book could be a two-brane of three dimensional space, for example. Another problem is how to decide which 4 dimensions are ours. There seems to be an infinite number of ways to combine them and the theorists want a unique way to arrive at our 4 and no other.

If all of this doesn't seem outlandish to you, then you are more brainwashed than you think. I don't have a problem with outlandish, *per se*. What I have a problem with is these guys claiming the moral and intellectual high ground of logic and rationalism. I also have a problem with them telling us there might be dimensions rolled up into space that we cannot detect when, all along, they have been telling us that if it can't be detected then it doesn't exist.

1. David Bohm, **Quantum Theory**, Prentice-Hall, Inc., 1951, Chapter 1, "The Origin of Quantum Theory."

2. See, for example, Nayfeh and Brussel, **Electricity and Magnetism**, John Wiley & Sons, 1985, Chapter 8.

3. Rupert Sheldrake, **Seven Experiments That Could Change the World: A Do-It-Yourself Guide to Revolutionary Science**, Park Street Press, 1995, Ch. 6.

4. Nick Herbert, **Quantum Reality: Beyond the New Physics**, Anchor, 1987.

5. Richard P. Feynmann, **QED, the Strange Theory of Light and Matter**, Princeton U. Press, 1985, p. 128.

Part III

A New Paradigm

7. Starting Over

"There is no shame in abandoning a path that has no heart" –Carlos Castaneda, **The Teachings of Don Juan**

I have spent my entire life trying to figure out what this physical world is, who I am and why I am here. After many, many years of thinking, studying, reading and meditating, I have come to the conclusion that our primary ideas about it are in error. These are our Sacred Cows and I am very aware that challenging them will likely not be appreciated by many. Because of this I hesitated to write this book at all. I also vacillated on writing this section as, no doubt, some will think me a quack. In the end I decided that it wasn't fair to tear down the Sacred Cows without offering an alternative view. My ideas about how physical reality works are tentative. Nothing has been worked out mathematically. As a result, this section may seem less precise than the previous ones, where many people have contributed over many years.

I am going to pretend that I am the czar of physics. The first thing I do as the czar of physics is to gather up every bit of information that I think may be pertinent.

1. Empty space is not empty.
2. Matter is not something.
3. Whirling motion is significant.
4. Physical reality comes in chunks (quanta).
5. Particles respond to changes in environment.
6. Something is funny about time.
7. Gravity is a problem.
8. Sometimes I lose things and they show up in places that I have already looked.

The next thing I will do as the czar of physics is to throw out ideas that do not fit. Nothing is sacred. The first thing to go is the Clockwork Rule (from Chapter 2). The Clockwork Rule was a knee-jerk reaction to a socio-political tyranny also known as the Holy Roman Church. It's time for us to get over it. The Clockwork Rule no longer serves us. The universe was not created out of nothing in the Big Bang and then left to its own devices (see Box 7.1). The order that exists in the universe could not have arisen by random chance from such a scenario. Yet, here we are.

Box 7.1 Rejection of the Big Bang
    I take exception to the Big Bang not only because it is inconceivable, but also because we really don't understand what time is. If time is artificial, then questions of beginnings and endings are naive and meaningless. The only evidence for the Big Bang that I

---

take seriously is the cosmological red shift. All of the other evidence is circumstantial. I must therefore find some other explanation for the observed red shift. I offer the following.

We know that light slows down in a medium by a factor of $1/n$, where n is the refractive index of the medium. The refractive index is dependent on the density and electrical properties of the medium. Along with a decrease in speed, there is a decrease in wavelength. When light exits the medium the speed resumes at light speed (c) and the associated wavelength returns to its value in vacuum. Throughout this process the energy, and thus the frequency ($E=hf$) remains unchanged.

The key point is this: it is the wavelength, and **not** the frequency that determines the resonance condition for absorption. Conventional wisdom has it that it is the energy (thus the frequency) that is important (*i.e.* the energy has to match the allowed energies available in the atom). I claim that it is the resonance condition that has to be met. And the resonance condition is dependent on the wavelength, not the energy, in analogy with the resonant modes on a string in Figure 6.1 and the arguments for the derivation of Plank's constant in Box 6.1. If this is true, then the wavelengths that are absorbed by the hydrogen in the atmosphere of the star will be associated with a lower energy (frequency) than those that would be absorbed by a dilute hydrogen gas because the waves inside the atmosphere have been shortened in proportion to the decrease in the speed of the wave as illustrated in the figure. When the radiation exits the star, the missing (absorbed) wavelengths will have been shifted to lower energy. The observed redshift in the absorption spectrum is then a measure of how compressed is the atmosphere of the source.

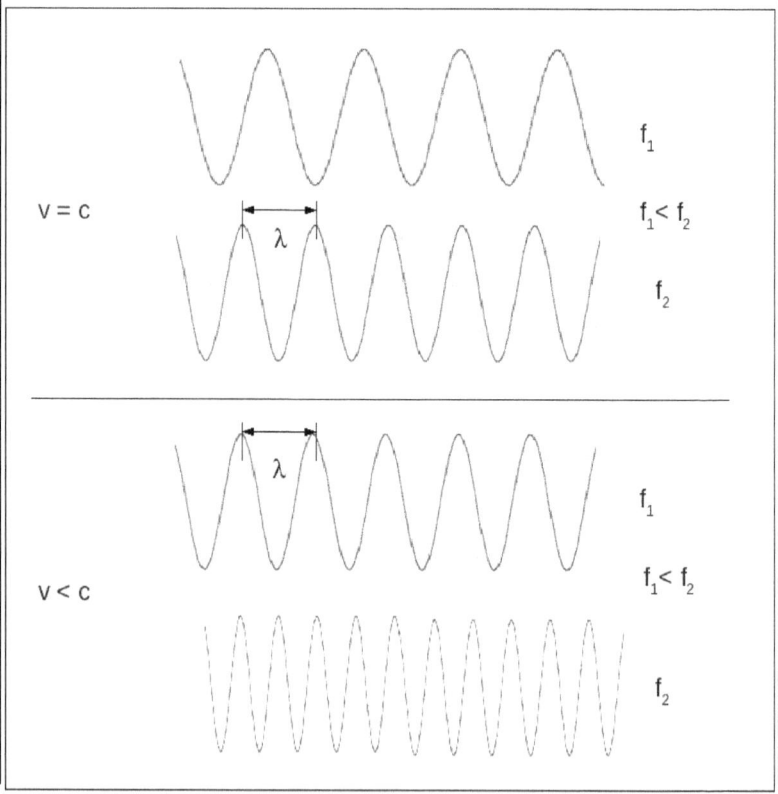

An astronomer named Halton Arp (b. 1924) has written books* showing data and explaining why the red shift cannot be caused by recessional velocity. "If Dr. Arp's earlier book, 'Quasars, Redshifts, and Controversies' put a few pinpricks into the Big Bang and Redshift-Distance Relation theories, this book blows open a hole so large you could drive a Mack truck through it."** In response, the scientific priesthood threw him off the 200-inch Hale telescope on Mt. Palomar and stopped publishing his papers. Business as usual.

* Halton Arp, **Quasars, Redshifts, and Controversies**, Interstellar Media, 1987.
** from a review of Halton Arp, **Seeing Red: Redshifts, Cosmology and Academic Science**, C. Roy Keys Inc., 1998.

As part of the Clockwork Rule, the idea that a phenomenon does not exist if we cannot detect it with our instruments also gets the heave-ho. This idea has already been seriously challenged by the infinite energy of "empty" space (the zero-point energy of the vacuum), virtual particle exchange, fields and string theories. So let's have done with it.

Now I can also dispense with the notion that the universe is a closed system. If all matter can only absorb and emit radiation in quanta of $E = hf$, then all radiation ever emitted in our universe can only have energy $E = nhf$, where n is an integer. "...even if some were present with other energies it could not, by hypothesis, interact with matter and, hence, would be undetectable."[1] There could be entire universes with slightly varying values for Plank's constant (h) that we wouldn't even know are there. In fact, there could be an infinite number of them. So the idea that we could just annex any unknown portions of the universe to our universe to make it a closed system is untenable. How will we know that we got them all if there are an infinite number of them and we can't even detect them? If our universe is not a closed system, then the second law of thermodynamics does not apply to the universe as a whole. There could be an exchange of energy between universes any time $nh_1f = mh_2f$ for universe #1 with h_1 and universe #2 with h_2, n and m are integers.

Before leaving the tossing out stage, I want to say a few words about time, though I don't understand it. According to Augustine of Hippo (a.k.a. Saint Augustine 354-430), time is a consequence of our consciousness. We humans have a limited ability to process information. In order to make the world compatible with our ability to observe differences we break the information into sequences (serial processing). The past is what we can remember. The future is what we are not able to remember. And the present is the transition point between past and future. Saint Augustine was trying to make an argument for the existence of God.

His point was that it is not difficult to imagine a being who does not have our limitation. A being capable of parallel processing does not exist in time. Saint Augustine called God's non-temporal state "NOW." NOW is not present. Now is a single state where all possible outcomes on everything that exists is available. I think of this like being on a big sphere, such as the Earth. Past and future already exist on the sphere, it's just that I have either already been there or I am not there yet. In either case they are not where I am now. Just like Italy does not exist for me where I am sitting now. But it is still over there on the other side of this sphere. I guess.

We still have time as a tag for the sequence of events. But time itself can not be observed. Not even quantum mechanics has a time operator. The only way we can measure time is by observing a periodic event and then count periods. We take an accepted or agreed upon value for the fundamental period and then count periods as needed. In physics, we use a continuous time axis (we think the axis is not continuous at the size of the space-time Plank length but even there the time intervals are uniform). A continuous time axis implies that every physical event no matter its nature evolves in such a way that I can always find a periodic event that SYNCHRONIZES with it. What fundamental principle is this based on?

I think that every physical configuration evolves consistent with its own internal clock and, in general, such clocks are not periodic. One consequence of this is that the fundamental unit of action $h/2\pi$ (Plank's constant) is not a universal constant but its value is adjusted such that the physical configuration can naturally evolve. In other words, within the evolution of a physical system $h/2\pi$ will be constant but the value of the constant may differ from the value used by a different configuration. If multiple realities exist, each with slightly differing values of Plank's constant, then we can switch realities by synchronizing or de-synchronizing our internal clocks.

Whatever time is, I don't think we need to be so strict about preserving causality in our theoretical formulations. If past and future do exist now, as Saint Augustine has said, then we should be able to "remember" the future as we do the past. And indeed, some of us do. We should also be able to change and affect the past, as we seem to be able to do with the future. One could also take the position that neither past nor future exist. We can't perceive them, measure them, or make any kind of case for their existence outside of our idea about them.

Now that I have summarily dismissed all ideas that don't seem to fit in with my list of observations, I have some room to maneuver. I want to introduce an observation that spirals and vortex motion are somehow significant. Spiral motion is stable as evidenced by spinning tops and gyroscopes, hurricanes, tornadoes, water spouts, galaxies, seashells, DNA structure, the solar system (see Box 7.2), orbital precession, spin precession and so forth.

Box 7.2. Vortex Motion

In Box 5.1, I made the case that our knowledge of the distance between the Earth and the sun was based on an *assumed* geometry. In this assumed geometry, the orbits of all the planets are more or less in the same plane. The axis of rotation of the Earth is tilted at an angle of 23.45 degrees to the plane of Earth's orbit about the sun. The Earth's orbit is slightly elliptical with an eccentricity of 0.0167. The geometry is shown in Figure 7.1a. I claim that, in order to be consistent with vortex motion, the orbit of the Earth about the sun is at an angle of 23.45 degrees and is precessing (as indeed, it does). This geometry is shown in Figure 7.1b. The orbits of the other planets would also have to be adjusted to fit this new geometry. I have asked some astronomers I know if we could tell the difference. They say they will get back to me, but they never have.

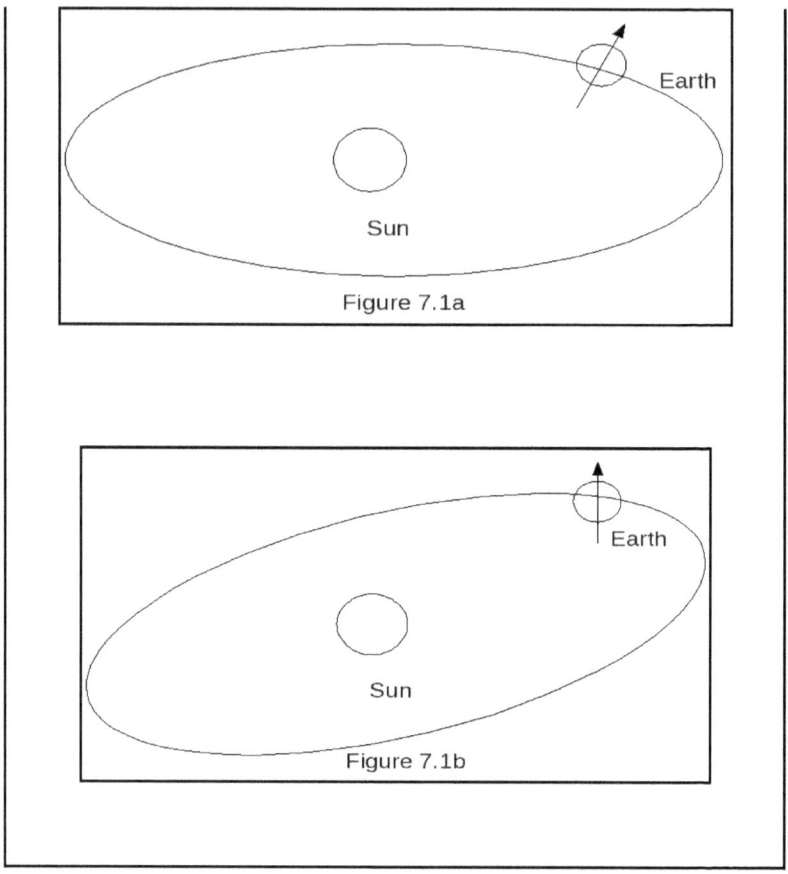

When I was taught about Einstein's equation $E=mc^2$, it was presented as if mass can be converted into energy and vice-versa. But it is more fundamental than that. Mass *is* energy. Suppose we have everything inside out. Suppose that space is something and particles are holes in space. Further suppose that space is infinitely dense, containing the possibilities and probabilities of everything that can exist. In order to remind myself not to take anything too seriously, I shall call this infinitely dense, perfectly elastic space the Pink Elephant Stuff (PES). In order to exist in the physical world, some-thing has to be differentiated from the every-thing contained in the PES. Suppose that holes are created in the PES by whirling energy, or a vortex. This vortex must push away the PES, creating a space for itself.

I would like to re-introduce Ann Conway's monads from Chapter 2 and identify them as the holes in the PES. Lady Conway imagined that the quasi-particles called monads were aware and that all matter was built up of various combinations of these bits of awareness in a great cooperative venture. Cooperation, and not competition, is the basis of existence. I do not mean to imply that the monads are like miniature humans running around communicating with each other. That would be very anthropocentric. I am saying that the monads are somehow aware of their surroundings and of the state of other monads. If they are aware of their surroundings and other quasi-particles, mustn't they also be self-aware? And is awareness and self-awareness not the definition of consciousness? Then these monads are alive. If the monads are alive, then so are elementary particles and everything else. You can't draw a line in the sand and put everything that isn't alive on one side and everything that is alive on the other. You cannot find such a line. If you can't find it, I claim it doesn't exist. This forces us to the logical conclusion that either every thing is alive or everything is not. Yet here I am.

The monads are far far beneath our perception or our detectability. Our elementary particles are built up of various combinations of these monads. The monads are of two types; one has a clockwise rotation and the other a counter-clockwise rotation. Energy flows into, through, and out of the monads. I shall designate the clockwise-rotation negative, with energy flowing from our universe to elsewhere; the counter-clockwise rotation positive, with energy flowing into our universe from elsewhere. Because energy is equivalent to mass, the positive monads thus have positive mass and the negative monads have negative mass. Various combinations of monads account for the mass, charge and spin that we measure in our elementary particles.

There have been two recent publications along these same lines. One, by Valery Chalidze (b. 1938), introduces the idea of vortices in a universal aether.[3] He calls these quasi-particles "vortons" and he relates the zero-point vacuum to a "vorton gas." Chalidze is synthesizing ideas from the 19th century originated by Michael Faraday (1791-1867), James Maxwell (1831-1879), William Thomson (1824-1907) and J.J. Thomson (1856-1940). Chalidze's universal aether is the same as my Pink Elephant Stuff. Though Chalidze equates mass with rotational energy of the vortons, he does not have negative mass, nor energy exchange. Further, he assumes that the universe somehow began with a huge vortex ring that spontaneously broke up into smaller and smaller rings until they met with stable conditions between the velocity of the rings and the universal aether.

In the second publication,[4] Friedwardt Winterberg (b. 1929) shows that all of the elementary particles can be built up from vortices of positive and negative mass quasi-particles that he calls Plank-mass particles. These quasi-particles exist on the scale of the Plank length (10^{-33} cm). Winterberg relates the Plank-mass particles to the zero-point vacuum energy and he calls it the Plank aether (See Box 7.3 about aether). There isn't a universal aether or PES in Winterberg's theory.

Winterberg claims that his theory preserves absolute space and time and it can all be done in four space-time dimensions, thus eliminating the need for all those messy extra dimensions of string theory. I don't have a dog in that fight. I like the idea that maybe my monads are rotating, closed strings. Perhaps even several entwined, rotating strings. The extra dimensions don't bother me because they can just occupy some other universe(s).

Box 7.3. The Aether

People have used the term aether to mean two entirely different things. One is the universal aether, or the Pink Elephant Stuff. The other is the aether as a medium for the transport of electro-magnetic energy (light). Most people think that the failure of Albert Michelson (1852-1931) and Edward Morley (1838-1923) to detect the aether in their famous experiment of 1887 killed the concept of aether as medium. In fact, Hendrik Lorentz (1853-1928) and Henri Poincare (1854-1912) developed the length contraction velocity transformations in an effort to preserve the aether. They showed that, because of length contraction, it was impossible to measure the aether using an apparatus such as Michelson & Morley used because the apparatus itself would be altered during the experiment. Einstein did not say the aether does not exist, he merely said we don't need it to explain light propagation as long as we assume that the speed of light is constant for all observers. He then used Lorentz's transformations as descriptions for his theory of Special Relativity. Einstein simply dodged the question of the existence of the aether.

The universe is populated by these quasi-particles of positive and negative mass that I call monads. I associate this monad sea with the zero-point energy of the vacuum, similar to Chalidze's vorton sea and Winterberg's Plank aether. When two or more monads combine, they also spin and rotate in another vortex, pushing out a larger space in the PES. This process is repeated, possibly many times, before finally forming protons, neutrons, and electrons, the building blocks that we know and love. The various combinations of monads determine the final "rest" mass and charge of the resulting "particles."

This process continues on up the scale to planets, solar systems and galaxies. Each new "no-thing" creates a disturbance in the PES. This disturbance is what Einstein called the warping of space in his theory of General Relativity (Chapter 4). As the "no-thing" called mass rotates and pushes out a new space in the PES, it not only causes a disturbance in the PES, but it also drags some of it around along the edges (called "frame dragging" in general relativity).

What looks like gravity is actually a pressure phenomenon. In Figure 7.2a, a single sphere in the PES experiences pressure from the PES pushing against it from all directions. When a second sphere is brought into close proximity, the pressure on sphere #1 is lessened due to the presence of sphere #2 (there is less PES in the "space" occupied by sphere #2) and the pressure on sphere #2 is lessened due to the presence of sphere #1. The two bodies will rotate about a common center of mass because of the disturbance each individually rotating body is creating in the PES. The stable motion for all matter is a spiral-type vortex with precession, like a spinning top. The warping of space that Einstein attributed to mass is the disturbance in the PES due to the vortex motion of confined energy, which *is* matter.

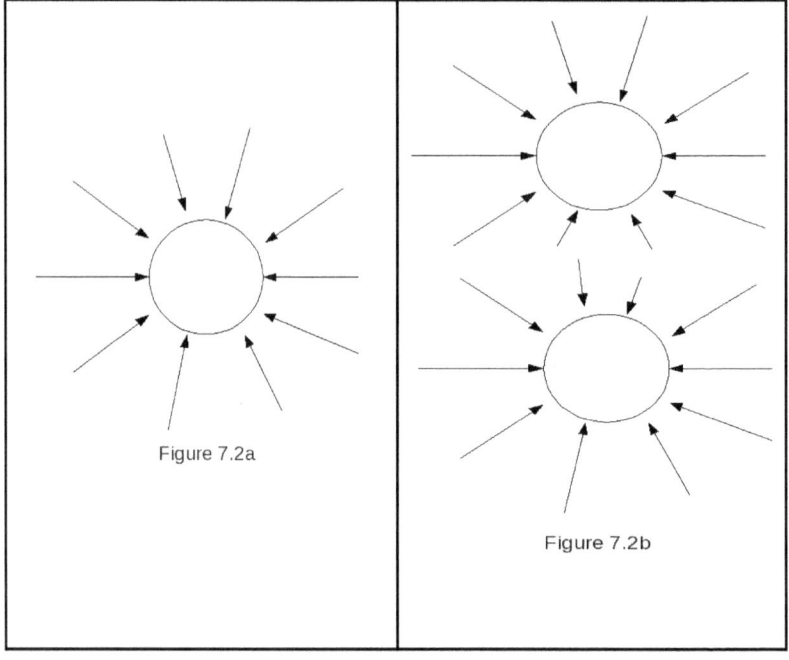

Figure 7.2a

Figure 7.2b

The universe, built up of monads, is alive and aware. Each "no-thing" has a combined gestalt awareness. All of the many monads that make up a proton know that they are part of the proton and the gestalt of proton monads is a new awareness of "protonness." Each proton, neutron, and electron that makes up the liver, for example, knows it is part of the "liver gestalt" as it performs "liver functions." The entire body has its gestalt and separate knowingness, apart from you. It has been shown that the very molecules in our bodies respond to our thoughts.[5] I once read about a guy who stepped into the street not noticing that a car was coming. His body jumped out of the way before he even knew he was in danger.

There is nothing "out there" but energy configurations. What we choose to perceive depends on our thoughts and our beliefs about what is possible. We perceive the energy that is in resonance with our thoughts (see Box 7.4). This explains the well-known phenomenon of people giving different versions of the same event.

You've heard people say, "we're on the same wavelength," or "I really resonate with that." These sayings are literally true. Nothing out there moves or changes. What changes is us. What changes is our thoughts and beliefs, which affect our clock synchronization and thus the reality that we perceive at any given moment.

We know from quantum mechanics that the physical world interacts with external energy only in chunks. This admits the possibility of entire universes that are off-resonance with ours. But, which one is "ours"?

I think that our world is discontinuous not only in space, but also in time. It blinks on and off. It does so at a rate that is outside our perception. Our equipment can't detect it because the equipment itself is blinking on and off at the same rate that we are. It is well-known that we don't perceive an event if it happens faster than a millisecond or so. That's why standard video is played at 24 or 30 frames per second. Anything faster than that is just a waste of bandwidth.

Suppose that we actually straddle realities. Our awareness can move smoothly from one reality to another by clock dis-synchronization. We automatically gloss over any anomalies, should we notice them at all. The sign that says, "Paris in the the spring," from Chapter 1 is an example. The missing car keys are another. We tell ourselves that the keys must have been there all along. But they weren't. This also explains why my memory of an event differs from, say, my Mother's. One of us has switched realities since then. And maybe switched back again. Probably many times.

Box 7.4. The Brain as Fourier Transform Processor

Our brains are Fourier Transform processors. What is "out there" is energy and patterns of energy. In order to perceive sound, for example, energy is incident on our eardrum as pressure waves. These pressure waves set our eardrums vibrating at so many vibrations per second in the time domain. Our brain processes that information and presents us with tones in the frequency domain.

Image-wise, what is "out there" is a series of energy patterns, say in line pairs per millimeter (spatial frequency – think plaid for simplicity). Our eyes focus light and dark interference patterns on the retina. Our brain processes the two-dimensional interference pattern and presents us with an image. How do you know that your image has anything to do with what is out there? How do you know that anything is out there?

Finally, let's review the various interpretations of the quantum mechanics measurement problem given in Chapter 6.

Interpretation #1: **The Copenhagen interpretation.**
Interpretation #2: **Observation creates reality**.
Interpretation #3: **Consciousness creates reality.**
Interpretation #4: **Hidden variables.**
Interpretation #5: **The many-worlds interpretation.**

Interpretation #6: **Quantum logic.**

I vote for all six. They are not at all mutually exclusive. What we need is a framework (paradigm) in which each of the six interpretations is one piece of the puzzle. From Interpretation #1, elementary particles are indeed not physical "things," they are bits of energy. They do exist independently of observation, but we can not know what they are apart from observation, leading to Interpretation #2. The energy pattern is perceived (measured) thus bringing them into the category of "thing," *i.e.* a physical entity with distinct properties. Interpretation #3 is satisfied by granting consciousness to everything (including elementary particles). This makes Interpretations #2 and #3 equivalent. Interpretation #4 is satisfied if we concede that when the particles are not physical objects being measured or perceived, they exist in the implicate order, or in some aspect of reality that we do not or cannot probe. Interpretation #5 says that there are an uncountable number of realities or "parallel universes" existing in the same "space." These are probable systems.

I have two small corrections to this interpretation. One is that reality does not branch every time a measurement (or choice) is made, but those realities exist now, and at all times. If two choices are possible and I choose one, my counterpart in a probable reality may experience the other *if that counterpart so chooses.* The vast field of probabilities exists always.

By our choices, we bring into our experiences the events that coincide with our desires, beliefs and expectations. This is the basis of free will. This also modifies our ideas about who and what we are.

The other correction is that we can and do have access to these probable realities, therefore they are not parallel. In fact, we can even switch realities. We do this all the time and don't notice because, as Everett claims, "the observer ... becomes correlated to the system." This also explains where my lost things go and why they show up in places I've already looked (Chapter 1). And why things I didn't know I had show up in my reality (the junk drawer). We don't notice because our consciousness jumps over any discontinuities. If we do happen to notice, we make up a story to explain any discrepancies.

Lastly, if everything, including elementary particles, has awareness, then the particles most likely follow some logic that escapes us (they are *not* miniature humans), which is in agreement with Interpretation #6.

We each are at the center of our own universe. We see only what we want to or expect to see out of all of the vast probabilities. All of the other actors are just script-holders in our play. Have fun with it.

1. David Bohm, **Quantum Theory**, Prentice-Hall, Inc., 1951, p. 20."

2. St. Augustine, **Confessions of St. Augustine,** Image Books Edition, NY, 1960.

3. Valery Chalidze, **Mass and Electric Charge in the Vortex Theory of Matter**, Universal Publishers, 2001.

4. Friedwardt Winterberg, **The Plank Aether Hypothesis; an Attempt for a Finitistic Non-Archimedean Theory of Elementary Particles**, Carl Friedrich Gauss Academy of Science Press, 2002.

5. Candace Pert, **Molecules of Emotion**, Simon and Schuster, 1999.

Epilogue

We don't know very much about anything and what we do know, we're not all that sure about. People who think otherwise have a position to defend because they have invested effort and they are confusing who they are with what they think or do.

Don't let anyone invalidate your experience. Don't let anyone tell you who or what you are. Do remember that they may not share in your experience and honor that. If you have a good idea, by all means, share it with us. But don't borrow bullshit scientific terms to explain your idea so it sounds more legitimate to you. It just sounds silly. Be authentic. Those guys don't know any more than you. Probably less.

To the scientific priesthood: If you want to keep your following, you guys need to offer up something more palatable than this lifeless, dry-as-dust reality you are currently stuck in. Stop invalidating people's experiences. We have all experienced the sensation of being stared at. Rupert Sheldrake made a valiant attempt at explaining such commonplace phenomena and got shot down for it. If you don't even try to incorporate people's experiences, either they will wander away one by one or they will stage a major revolt. Who do you think funds your research?

There are only three reasons I can think of why people still worship you.

1. They like their gadgets. They haven't yet noticed that their gadgets only give the illusion of happiness while distracting them from their pain. Like any addiction.

2. Their babies stopped dying of polio, diphtheria, tetanus and so on. But the generation who remembers that and is grateful is dying off.

3. You threaten them with ignorance, weeping, gnashing of teeth, a return to superstition and upheaval if they don't listen to you. You (of all people) accuse them of being irrational if they dare to question you.

I like my electric gadgets as much as anyone and I don't wish to give them up. However, I would far rather wade through sheep shit all day than sit in a plastic cube staring at a computer screen. Friends, the Dark Ages are upon us.